PHOTOSENSITIVE GLASS AND GLASS-CERAMICS

PHOTOSENSITIVE GLASS AND GLASS-CERAMICS

Nicholas F. Borrelli
Corning Inc., Corning, New York, USA

CRC Press
Taylor & Francis Group
Boca Raton London New York

CRC Press is an imprint of the
Taylor & Francis Group, an **informa** business

CRC Press
Taylor & Francis Group
6000 Broken Sound Parkway NW, Suite 300
Boca Raton, FL 33487-2742

First issued in paperback 2019

ISBN-13: 978-1-4987-4569-7 (hbk)
ISBN-13: 978-0-367-87512-1 (pbk)

Library of Congress Cataloging-in-Publication Data

Names: Borrelli, Nicholas F., 1936- author.
Title: Photosensitive glass and glass-ceramics / Nicholas F. Borrelli.
Description: Boca Raton : Taylor & Francis, a CRC title, part of the Taylor & Francis imprint, a member of the Taylor & Francis Group, the academic division of T&F Informa, plc, [2017] | Includes bibliographical references and index.
Identifiers: LCCN 2016001664 | ISBN 9781498745697 (alk. paper)
Subjects: LCSH: Glass, Photosensitive. | Glass--Optical properties. | Glass-ceramics--Optical properties.
Classification: LCC TP862 .B67 2016 | DDC 666/.156--dc23
LC record available at http://lccn.loc.gov/2016001664

Visit the Taylor & Francis Web site at
http://www.taylorandfrancis.com

and the CRC Press Web site at
http://www.crcpress.com

To the memory of the two research pioneers Wm. H. Armistead and S. Donald Stookey, who through their efforts initially discovered many of the amazing materials described herein.

Contents

Preface

One could make the argument that during the last 100 years more new technological advances have been made than in the history of humankind until this point in time. Clearly, an important part of this is due to the tremendous advancements in the physical sciences, but also an equally important part has come from the concomitant rise of industrial research laboratories throughout this country. These laboratories, from Edison's early General Electric Lab, to the more present-day Bell Labs, transformed these new scientific discoveries into useful and enabling products. You need only look at the smartphone in your pocket to appreciate the progress that has evolved. The Corning Research Laboratory was one of those early labs and sadly is one of only a handful that has survived. This book to some extent chronicles part of its contribution to this unique period in time.

Nicholas F. Borrelli
Corporate Research Fellow
S&T Division
Corning Incorporated
Corning, New York

Acknowledgments

I would like to thank D.L. Morse and G.S. Calabrese for their support in allowing me to pursue this effort. Special thanks are given to R.J. Araujo and T.P. Seward, III, and to George Beall, whose contributions and discussions could very well warrant them to be coauthors of this book. The contribution and support of my coworkers, both past and present, are gratefully acknowledged since this book is as much theirs as mine. I would specifically like to thank Joe Schroeder for his tireless work on the graphs; Charlene Smith, Matt Dejneka, Wageesha Senaratne, Nadja Lonnroth, and Angela Law for their help; and others who developed many of the measurements that appear in the book. I also thank Ashley Weinstein and Ashley Gasque of CRC Press/Taylor & Francis for their guidance and support. And last but certainly not least, thanks to my beautiful wife Kaye and our entire family for their continuing love and support.

And finally, I would be remiss not to remember all the members of the Corning Incorporated Research Laboratory, from day one to the present—all 108 years—for making this lab what it is today.

Author

Nicolas D. Borrelli, PhD, received an MS (1960) and a PhD (1963) in engineering from the University of Rochester (New York). His thesis was on the interpretation of the IR spectra of borate glasses. He started at Corning in 1962 as a research scientist. His first project involved the study of Nd:Glass Lasers. Corning was the second laboratory in the world to demonstrate laser action in glass. He subsequently worked on early aspects of laser communication, such as optical modulators and optical isolators. Dr. Borrelli also worked on projects related to the first optical fiber, which was being developed at Corning at that time. He later moved to the study of photochromic and photosensitive glass, which led to two Corning products: a polarizing glass called Polarcor™ and a Serengeti™ sunglass. His interest then shifted to the new field of microoptics, which later led to a book entitled *Microoptics Technology* in two editions (Marcel Dekker 2005), and a microoptic lens array product, SMILE™. Dr. Borrelli has published widely, more than 150 papers in technical journals and four contributed book chapters that cover areas involving nonlinear optics of glass, photochromic and photosensitive glasses, and quantum dots in glass. His current interests range from antimicrobial glass to novel optical properties of glass-ceramics with a renewed interest in photochromic glasses. Dr. Borrelli holds 145 U.S. Patents, which cover many different topics. He recently received the President's Award from the International Commission on Glass given for lifetime achievement in glass science.

1 The Glass Menagerie

Title of a Tennessee Williams Play (1944)

1.1 INTRODUCTION

Glass is normally thought of as an inert material, one that is stable to the ambient conditions and will remain so for eons. This property of inertness plays a dominant role in glass's numerous applications ranging from a simple window to the modern complex optical waveguide that drives today's optical network. However, there is a special class of much less known glass that can be made to interact strongly with its environment, and most importantly, with light either in a transient or permanent way. The general classification is termed "photosensitive," which actually pertains to a number of different manifestations of the phenomenon although they contain a common physical basis. The simplest case is solarization, where the optical effect is observed after exposure. An example of photosensitive glass is one that colors when exposed to long periods of sunlight, such as glasses often used in the desert, and hence the origin of the term solarization. This property derives, as we will see, from a number of different sources; for example, from small amounts of naturally occurring impurities retained as the glass was formed.

The second effect, is termed photosensitive, is where an optical effect is seen only after a subsequent thermal treatment to the exposure. The third category and likely the most familiar example of a transient response would be a glass that darkens in response to sunlight and fades in the shade. This property is termed somewhat inappropriately as "photochromic" since there is no color change involved, just a uniform darkening. In this special case the glass has additional impurities (dopants) added in the melting stage that ultimately accounts for the photochromic effect through the development of an Ag halide phase that is very similar to what is used in photography. This will be discussed extensively in Chapter 7.

The photosensitive version to be covered in this book is of the permanent variety where special glasses that are exposed to light in an initial state then subsequently heated or otherwise treated develop a permanent color (Chapter 2) or a change in refractive index (Chapter 4). The most interesting effect is when a microcrystalline phase within the glass is nucleated utilizing the photosensitive mechanism. In other words, the color-producing agent itself can initiate the growth of a nanocrystalline phase from the glass. This latter phenomenon was pioneered by research at Corning Incorporated in the 1950s that later was commercialized into such products as Fotoform™, Fota-Lite™, and Polychromatic™ (Chapter 3). Many of the examples of this phenomenon are closely related in mechanism to the historical conventional film-based photographic process. In fact, the discovery of many of the physical manifestations mentioned above stemmed from the early work of Corning's S. Donald Stookey[1] who was trying to make glass into a direct photographic medium and not

1

just the substrate for a photographic emulsion. This will become more evident in Chapter 2, when the physics and chemistry of the photosensitive mechanism are discussed.

Until now, the role of the glass composition, and even more to the point, the glass structure as it pertains to the propensity of the glass to exhibit photosensitive behavior, has not been mentioned. By glass structure, we mean the physical atomic structure of the glassy network of atoms. Because glass is often described as amorphous, this is somewhat misleading. Although glass has no regular long-range order as do crystals, it does have short-range (regular silica SiO_4 tetrahedra) and quasi-ordered intermediate showing up as quasi-ordered N-membered rings. This is schematically shown for silica in Figure 1.1. This could be viewed as a disordered crystal where the disorder is in the long-range distances.

However, there are other properties which are quite similar to crystals that are not significantly dependent on an ordered lattice of atoms. Perhaps surprisingly, the electronic band structure is similar. Figure 1.2, from Sigel,[2] shows the measured deep UV spectrum comparing silica to its crystalline counterpart, quartz.

The spectral comparison essentially shows that the energy levels available to the electrons as expressed by the electronic band structure are very similar. In a way this is not really so surprising when one realizes that the electronic energy levels arise from the atomic orbitals, mainly p-orbitals of the oxygen (valence band) and sp3 orbitals of the silicon in silica that makes up the conduction band.[3,4] However the disorder does have an effect in producing additional localized states above the valence band and below the conduction band. These localized states are shown in Figure 1.3.

One now defines a mobility gap because electrons promoted to these states cannot contribute to electronic conduction because the states are localized. Electronic conduction occurs through electrons being able to move continuous bands formed by the overlap of the atomic orbitals. The electrons must occupy the higher extended

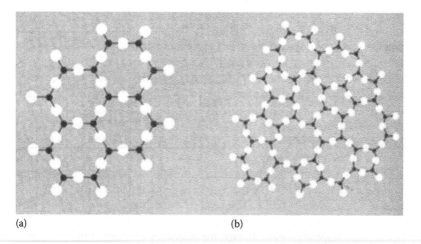

(a)   (b)

FIGURE 1.1   A schematic two-dimensional representation of the difference between a crystalline structure (a) and its analogous disordered or glass counterpart (b).

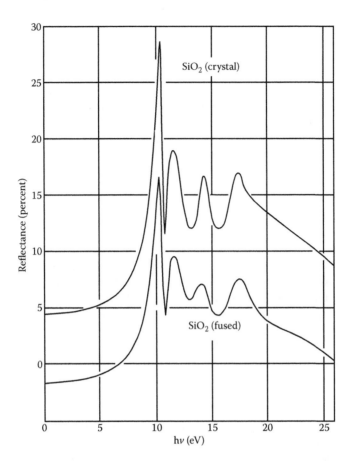

FIGURE 1.2 Vacuum UV spectrum of crystalline quartz and fused silica showing the remarkable similarity of the position of the energy bands. (From G.H. Sigel, *J. Phys. Chem. Solids* 32, 2373, 1971.[2])

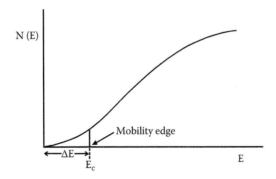

FIGURE 1.3 Schematic representation of the density of electronic states for a disordered structure such as silica showing the existence of a localized state and extended states, the latter is able to carry electrons.

continuous states in order to be mobile enough to carry current that then determines the real bandgap as would be defined in a crystal.[5,6] However, localization of electronic states is always the consequence of disorder, as Anderson first asserted.[7] As we will see, this aspect of the electronic structure plays a large role in understanding the photosensitive process. We will often refer to defect or impurity states as the source of the photosensitive effect. It is clear that the understanding of the physics and chemistry of oxide glasses and the amorphous state in general is an ongoing quest and further aspects such as photosensitive effects just adds another layer of complexity to an already complex situation. Nonetheless, one still tries to be as quantitative as possible in the explanations of the various photosensitive effects that our present understanding of the glassy state allows. This lack of understanding particularly applies to the electronic structure of glass, which then plays a significant role in the photosensitive mechanism, as we will see in the next chapter. Surprisingly, there is a misunderstanding by many in the field about the electronic structure of disordered materials as we discussed above, which hampers the progress of many of the issues we cover in this book. Glass has an electronic band structure as do all solids it is just not readily described by the mathematical framework that has been built so well for crystalline materials, therefore making it more difficult to produce quantitative predictions. This all notwithstanding, we proceed on.

In the above discussion the phrase *special glasses* was used, which implies that not all glasses can be made to be photosensitive; that is, to be altered in some way by light. This is generally true in all but one or two of the topics covered in this book. However, most of the photosensitive glasses all are members of the general class of common glasses ranging from soda-lime to alkali-aluminoborosilicates. Moreover, the role of the glass composition differs significantly depending on the specific photosensitive effect in question (e.g., changes in color, refractive index, or photochromism). As we will see, the specific photosensitivity derives from the ability for the structure to include dopants such as halogens or silver ions or the presence of inherent structural features such as nonbridging oxygen. These latter features supply the defect sites that trap charged species that are created by the light. This will be made clear as we discuss each topic in detail.

The individual chapters will deal with the specific photosensitive phenomena after a review of the basic photosensitive process in Chapter 2. Chapter 3 explains the photonucleation phenomenon where the initial photosensitive step initiates production of a crystalline phase. Chapter 4 discusses the consequences resulting from the change in the refractive index produced by these photonucleated phases. Chapter 5 reviews photochromic glasses followed by a discussion in Chapter 6 of an interesting aspect of these glasses called photoadaption, where light is now seen to interact with the photochromic phase. Chapter 7 considers the phenomenon of solarization where induced coloration is produced in glasses by UV light. This is one area in which all glasses experience the effect to some degree or another; in other words, these are not special glasses in the sense that nothing is intentionally added to produce the photosensitive effect. In Chapter 8 we discuss a variation on the photosensitive theme by presenting a number of interesting photochemical and photophysical effects in a special glass host. In Chapter 9, the discussion focuses on how special glasses combined with glass-forming processes can be made into efficient polarizers. Although strictly speaking, these are not photosensitive glasses as defined above, the effects have much in common. In fact, one of the glasses utilized is derived from a photochromic glass which is described in Chapter 5.

Overall, the range of effects described and discussed throughout this book are as broad and diverse as the properties of glass itself, but this is not to say that they are not dealt with in as detailed a manner as possible. In a recent book by Steven Johnson,[8] *How We Got to Now*, glass is described by its unique and ubiquitous presence in our technological world over the centuries. This present book in some small way reinforces how this may have come about.

REFERENCES

1. S.D. Stookey, *Journey to the Center of the Crystal Ball*, American Ceramic Society, Columbus, OH, 1985.
2. G.H. Sigel, *J. Phys. Chem. Solids* 32, 2373, 1971.
3. M.H. Reilly, *J. Phys. Chem. Solids* 31, 1041, 1970.
4. A.R. Ruffa, *J. Non-Cryst. Solids* 13, 37, 1973/74.
5. N.F. Mott, Electrons on glass, *Nobel Lectures in Physics 1971–1980*, World Publishing, Singapore, 1982.
6. N.F. Mott and W.D. Twose, *Adv. Phys.* 10(38), 403, 1961.
7. P.W. Anderson, *Phys. Rev.* 109, 1492, 1958.
8. S. Johnson, *How We Got to Now*, Penguin Press, 2015.

2 Photosensitive Glasses

A picture is worth a thousand words.

Chinese proverb

2.1 GENERAL INTRODUCTION

Photosensitive glasses that produce color upon UV irradiation followed by thermal treatment have been known for many years.[1-3] A typical example of a photosensitive image in glass is the so-called gold ruby as shown in Figure 2.1. The "ruby" notation comes from the color and is actually produced from gold nanoparticles, as will be shown. A typical photosensitive glass composition is provided in Table 2.1.

The color arises from the surface plasmon resonance absorption (see Section 2.5.2) of the gold nanoparticles that were produced by the exposure to ultraviolet (UV) light followed by a thermal treatment at 550°C. In general as in the Au example, the glasses contain a photosensitizer (that produces photoelectrons), usually Ce^{+3}, and in addition a noble metal ion such as Ag or Au which will eventually combine with the electron to produce metal nanoparticles after thermal treatment. Other photosensitizers have been demonstrated, such as Eu^{+2} and Cu^{+1}. We will discuss the use of Cu^{+1} in more detail in Section 2.4.2. Also note that the use of a photosensitizer is not necessary when the energy of the exposure source is larger than the electronic bandgap of the glass (see Chapter 1) in such cases as using a 254-nm excimer laser or a 800-nm fs-laser, the latter being effective through a multiphoton effect because of the high laser pulse intensity.[4] In these situations, the electron is promoted directly from the valence band to the conduction band of the base glass. These latter points notwithstanding, a simple reaction representation of the process is given by the following.[3]

$$Ce^{+3} + h\nu \rightarrow e + Ce^{+3+} \ (250 - 350 \text{ nm})$$

$$Ag^+ + e + \text{heat} \rightarrow Ag_0$$

$$n(Ag_0) + \text{heat} \rightarrow (Ag_0)_n \tag{2.1}$$

There are more steps involved than this simple picture depicts in that the holes and electrons produced by the light must be localized to some extent by some sort of trapping site, which are involved in the process, otherwise recombination would dominate.[5] An example of such a site is suggested by some experimental results that could indicate that only glasses with nonbridging oxygen are seen to be photosensitive, pointing to role of the NBO as the hole trap[5]—otherwise, the photoelectron could easily recombine. Similarly, the electron is likely initially trapped and

FIGURE 2.1   An example of an image in gold photosensitive glass.

TABLE 2.1
Typical Photosensitive Glass Composition

Compound	Weight (%)
SiO_2	75.23
Li_2O	9.45
Al_2O_3	9.82
Na_2O	1.22
K_2O	2.37
Sb_2O_3	0.022
SnO_2	0
CeO_2	0.01
ZnO	1.6
Ag	0.08
Au	0

subsequently released as the temperature is raised in the heat treatment stage, at which point it combines with the diffusing Ag ion and becomes more mobile because of the elevated temperature and hence becomes more likely to come in contact with the electron. This conjecture is supported by the experimental results that if after exposure the glass is held at ~200°C for a prolonged time prior to the standard thermal development schedule (~500°C–600°C), the photosensitive effect is quenched; that is, no color is produced. This implies that the electrons can be thermalized back

into the conduction band at a lower temperature without combining with the Ag ion
that is not yet mobile enough at the somewhat lower temperature. In other words,
they cannot diffuse rapidly enough to meet up with the short mean free path electron
now traveling in the conduction band. In view of this more detailed view, the above
equations expressing the phenomenon should be augmented.

$$Ce^{+3} + hv \rightarrow e + Ce^{+3+} \ (250 - 350 \ nm) \tag{2.2}$$

$$e + T_e \leftrightarrow T_e^e \tag{2.3}$$

$$Ce^{+4} + T_h \leftrightarrow T_h^h + Ce^{+3} \tag{2.4}$$

$$Ag^+ + e + heat \rightarrow Ag_0 \tag{2.5}$$

$$n(Ag_0) + heat \rightarrow (Ag_0)_n \tag{2.6}$$

The inclusion of Equation 2.6 accounts for the possibility of the hole being ultimately
trapped elsewhere. In a set of absorption curves beginning with Figure 2.2a, the Ce^{+3}
absorption results from the photoexcitation step in Equation 2.2 are shown; Figure 2.2b
shows the spectrum of the induced absorption from the production of trapped electron
and hole states represented by Equations 2.3 and 2.4, and Figure 2.2c. The TEM image
of the Ag speck is represented as the silver forming steps in Equations 2.5 and 2.6.

There is an interesting phenomenon in Figure 2.2d, which is linked to the growth
of the Ag by showing the fluorescence as a function of the heating attributed to the
Ag clusters. Some researchers have conjectured that the electron trap may very well
be some sort of Ag_n^+ species, as we show in Figure 2.2d. The fluorescence is attributed to the radiative recombination of electrons with the trapped holes. The competi-
tion then is between the recombination and the Ag_n^0 formation. It has been shown
that if one uses a thermal cycle that heats too rapidly to ~500°C, one can favor the
recombination and frustrate silver formation.

$$T_e + heat \rightarrow T + e$$
$$e + Ag_n^+ \rightarrow Ag_n^0 \tag{2.7}$$
$$e + T_h' \rightarrow hv$$

The third part of Equation 2.7 indicates the fluorescence phenomenon referred to
above. In one study, Maurer[6] used light scattering to follow the growth of an Ag
particle though the thermal development stage.

The theory of the photographic process deals with the phenomenon in some detail
in the case of Ag halides; this will be covered in more detail in Chapter 5, which

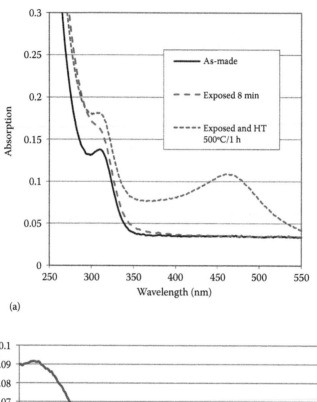

(a)

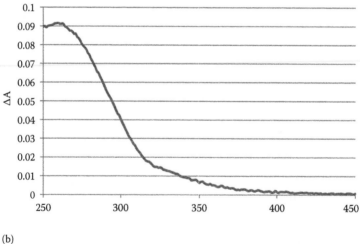

(b)

FIGURE 2.2   Relating to the steps in Equations 2.2 through 2.6: (a) the absorption curve showing the Ce^{+3} photoexcitation step corresponding to Equation 2.2, (b) the induced absorption corresponding to the trapped species corresponding to Equations 2.3 and 2.4.

(Continued)

(c)

(d)

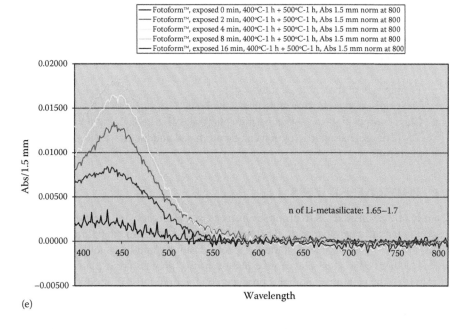

(e)

FIGURE 2.2 (CONTINUED) Relating to the steps in Equations 2.2 through 2.6: (c) TEM image of Ag-speck as represented in Equation 2.6, (d) fluorescence attributed with Ag-ion clusters (see text), and (e) surface plasmon absorption of Ag-speck indicating the increasing number with higher development temperature.

discusses photochromic glass. In the latter case, it has more relevance in that it deals with the photolysis of the Ag halide nanocrystals. Nonetheless, one might expect that formation of Ag nanoparticles in glass by a photolytic process might have something in common with the formation of the Ag-particle in photographic emulsion. Mott and Gurney's[7] theory posits that more electrons are trapped at the growing Ag cluster, charging it negatively and thus attracting more nearby Ag ions to the particle. If this is true and it is not unique to the host being a halide crystal, then it might also occur in a disordered structure. Initially, the absorption of Ce^{+3} at 310 nm of the unexposed glass is seen, which is followed by the spectrum after exposure, and finally the spectrum after the thermal treatment showing the Ag nanoparticle surface plasmon resonance at 460 nm (refer to Chapter 6 for a further discussion of surface plasmon behavior). Figure 2.2e shows the increase in the number density of Ag nanoparticles as a function of time of heating at 500°C after exposure. The induced absorption attributed to all the trapped species other than those that are directly involved in the photosensitive process as a result of the exposure (see Chapter 7) is also seen. Note that after the thermal treatment the absorption peak from the Ce^{+3} seems to be recovered to a large extent, which indicates that Equation 2.6 is playing a role. In other words, the hole is not ultimately trapped on the Ce^{+3}. This fact is consistent with the observation that the amount of silver produced is not a function of the initial Ce^{+3} content, which indicates that the hole is eventually trapped elsewhere. This means the Ce^{+3} is kind of a pump that produces electrons and at the same time is also receiving electrons.

The following sections will discuss each of the processes involved in Equation 2.2 as listed below:

- The physics of the photoexcitation step
- The origin of the electron/hole traps involved
- Glass composition space that permits photosensitive behavior and examples
- The different nanoparticles that can be produced and the absorption derived therefrom

2.2   PHOTOSENSITIZERS

The role of the sensitizer is to produce electrons that ultimately reduce metal to form absorbing metallic nanoparticles. This will be discussed in some detail in Chapter 7 where the more general manifestation of photosensitivity and the role the electronic band structure of the glass plays in the photosensitive process are explained. Suffice it to say, that the electrons and holes produced by the light must be mobile enough to combine with metal ions. The only possible way this can happen is that through the absorption of the photosensitizer, electrons are promoted to the extended electronic states referred to as the conduction and valence bands of the glass (see Chapter 1). As a result, the role of a photosensitizer ion such as Ce^{+3} is that somehow through its optical excitation it can promote an electron into the conduction band of the glass. In general both the ground and excited states of the impurity ion in the glass would lie within the forbidden gap as defined in Chapter 1. Further the optical transitions of the rare-earth impurities sensitizers depicted in Figure 2.4 are those involving 4f-5d levels and not the common inner 4f-4f transitions. These are weak transitions because they are parity-forbidden ($\Delta L \neq 0$) to the first order. However, there are also charge transfer transitions where the electron is transferred to the oxygen ion ligand that is bonded to the metal ion. These transitions are strong because they are parity-allowed. It is from these types of transitions that photoexcitation to the conduction may occur since the p-orbitals of the oxygen and p-orbitals of the silicon make up the bands of the electronic structure of the glass. The picture that emerges is shown in Figure 2.3.

To be a photosensitizer, the impurity ion energy levels must lie in the gap in such a manner that the energy of the charge transfer level lies above the conduction band, as shown in Figure 2.3a. The three known photosensitizer ions shown in Figure 2.4 all satisfy these conditions; that is, their positions in the gap are such that their charge transfer levels are above the conduction band of the glass. Further, to complete the argument, one has to conjecture the orbitals making up the excited state levels of the photosensitizer ion can admix with the orbitals making up the conduction band states, thus allowing the probability that the electron can end up in the conduction band.

In the case of Cu^{+1} as the photosensitizer is likely the same, the same mechanism and the location of the ground state within the bandgap can be estimated as 320 nm from the absorption cutoff shown in Figure 2.5 for a glass with ~0.6% Cu all as Cu^{+1}. Therefore, for all three known photosensitizers the excitation wavelengths are in the 320-nm region.

As mentioned above one does not need a photosensitizer dopant if the energy of the light source is sufficient to bridge the gap of ~250 nm. The convenience of having an excitation source in the 320–350-nm range is great since most standard photolithographic capability exists in this range. The typical exposure using an Hg or HgXe lamp operating at 320 nm is of the order of 10 mJ/cm^2 for 10–20 minutes (~10 J/cm^2) exposed through a 1-mm thick sample.

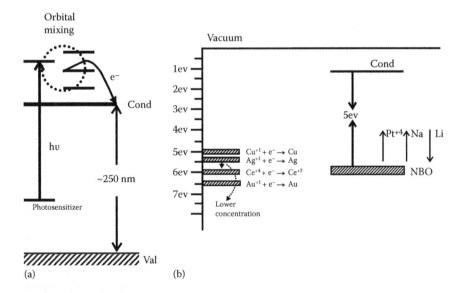

FIGURE 2.3 (a) Schematic energy diagram indicating the photoexcitation process involving the Ce^{+3} photoelectron ending up in the conduction band of the glass, and (b) a more detailed energy diagram showing the energy position of the respective photosensitive ions.

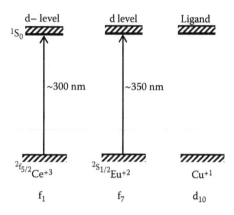

FIGURE 2.4 Energy levels of the three known photosensitizers indicating their excited state overlap with the conduction band of the glass.

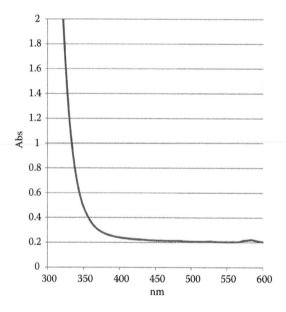

FIGURE 2.5 Absorption cutoff of the Cu$^+$ ion in photosensitive glass.

2.2.1 A More Detailed Mechanism

There are some needed refinements of the photosensitive mechanism given above that explain some experimental results and also bear on some aspects of the results in the next chapter when we deal with phases nucleated by the noble metal nanoparticles. It is appropriate to deal with them here at this point in the discussion of the initial mechanism. There are two experimental results not explained by Figure 2.3a. First, we have found that that if the alkali ion in the glass is all Li rather than Na with all other components of the glass being the same, then there is no evidence of a photosensitive effect. The second result is that when the alkali in the Na glasses and one lowers the Ag content in the absence of Au the photosensitive effect disappears and then reappears at a level 20 times that of the lower level. To explain these effects, it is important to appeal to a more exact energy diagram, as shown in Figure 2.3b. What is plotted is the potential in eV for the reduction reactions indicated. It also shows the position of the NBO level estimated from the work function and the energy gap.

The argument follows the steps below based on Figure 2.3:

1. The NBO level lies above the Ce^{+4} energy, indicating the trapping of the hole by the NBO.
2. If the NBO level lies below the Ag^{+1} level the electron is attracted more to the NBO than to the Ag^{+1}, and therefore there is no reduction. This is the case for Li glass. For Na glass the bandgap is smaller and the NBO level is above the Ag^+, and therefore reduction occurs.
3. The Au^{+1} level always lies *below the NBO level* so the more stable condition is that the hole remains on the NBO and the Au^{+1} is reduced.

Based on these steps, the conclusion would be that Au^+ is the more efficient electron trap irrespective of the kind of alkali present. The energy diagram Figure 2.3b could also explain why in the Li-containing glasses the Au^+ is not required if the Ag concentration is low enough. In this case one can turn to the Nernst equation, which states that the chemical potential of a silver reduction reaction depends on the concentration.

$$E = E_0 - 0.059 \log \frac{Ag}{Ag^{+1}} \qquad (2.8)$$

Applying the ratio of the Ag^{+1} concentrations for the high- and low-silver glass from the above set, one could lower the Ag reduction potential by as much as 0.1 eV, putting it below the NBO level.

2.3 ELECTRON AND HOLE TRAPPING

As mentioned above, additional discussion of this topic is presented in Chapter 7, which deals with solarization. Solarization, which is the induced absorption upon exposure to UV light, is a more general and ubiquitous version of the generic photosensitive process and although it shares many common elements it nonetheless is distinguished from the word photosensitive in the following ways:

- Solarization requires no sensitizer
- All glass solarizes to some extent whereas only special glass compositions can be made photosensitive
- Absorption is observed directly from exposure
- Solarization does not involve the formation of a nanophase either directly or by subsequent heating
- Induced absorption can be reduced with heating while photosensitive absorption is permanent

In fact, some degree of solarization is usually observed after the initial exposure of the photosensitive glass but this disappears during the subsequent thermal development schedule that produces the permanent absorption. What the two phenomena share in common is that they both require the production of photoelectrons, holes, and energy levels where they become subsequently trapped. In the case of solarization, the trapped species are the source of the induced absorption while in the photosensitive glass the trapped electrons and holes play an intermediate role leading to the induced absorption. In other words, they react with metal ions to form the absorbing species. The nature of the traps are most likely different in the sense that for the photosensitive process the traps must be much deeper because there is a high temperature step that anneals out the normal solarization. In other words, the traps have to be deep enough so that the electrons (holes) are not thermalized until the metal ion has sufficient mobility to move through the glass network. It is thought that only one trap has to be deep if one assumes that the unwanted recombination is always between the electron and the hole in the respective bands and not between the electron and a trapped hole or a hole and a trapped electron. In the listing of differences above we made the point that not all glass compositions can be made to produce the photosensitive effect by adding a photosensitizer and a noble metal ion. This classification of glass that can be photosensitive needs the presence of nonbridging oxygen as a critical requirement. The proposed deep trap for the photosensitive process is the hole trapped on the nonbridging oxygen.[5] A schematic of a plausible energy diagram of the trapping of the hole by a localized level of the valence band is shown in Figure 2.6.

The hole trap must be deep enough (far enough above the valence band) so that the holes are not substantially released until the temperature is high enough for the Ag ions to have sufficient mobility through the glass. This assumes that the probability of the formation of the Ag^0 requires both electrons in the conduction band and the Ag^{+1} ions' mobility throughout the glass network. We will see how this hypothesis leads into the next section when the role of glass composition as it pertains to this photosensitive property is investigated.

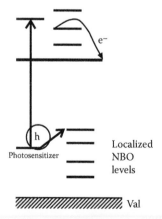

FIGURE 2.6 Schematic energy diagram indicating the trapping of a hole by a localized level of the valence band.

2.4 PHOTOSENSITIVE GLASS COMPOSITIONS

2.4.1 NOBLE METAL-BASED

Up to this point in the alkali aluminoborosilicate silicate system the only oxide glasses that have been shown to form the photosensitive noble metal nanophase are those that contain nonbridging oxygen. There are some generally accepted structure-based rules for the existence of NBOs. In the alkali aluminoborosilicate system the rule is simply that if the total alkali content on a molar basis exceeds the alumina content the NBOs will be present proportional to the disparity of the number. This rule is a result of the simple consequence of the 4-coordinated Al maintaining a totally connected network that has no breaks corresponding to an NBO. If there is excess alkali, then NBOs are required. This is shown in Figure 2.7a.

For the alkali aluminosilicate case this has been experimentally verified. Glass compositions listed in Table 2.2 have been shown to be photosensitive when the alkali $\gg$ alumina and not when alkali was $\leq$ unity.

This rule extends to the alkali aluminoborosilicate system as long as the borons are all in 4-coordination with the oxygen. In the alkali borosilicates, the problem is more complicated in that the boron coordination varies depending on the alkali/boron ratio. This is shown in Figure 2.8 where the ratio of 4-coordinated to 3-coordinated borons is plotted against the alkali/boron ratio. All the boron is in 4-coordination up to a certain value of the alkali/boron ratio, and hence there are no NBOs, but when this value exceeds a certain number depending on the silica/boron ratio, then NBOs start to appear.

For the alkali borosilicates we used the graph in Figure 2.8 to select four alkali borosilicate glass compositions with Ce/Ag added where two are taken from the region of the graph and where all the boron is 4-coordinated (alkali/boron = 0.3 and 0.63), and two compositions taken from the right side of the break in the curve (alkali/boron = 1 and 1.44) where now some of the borons are 3-coordinated and hence have NBOs. The silica/boron ratio was fixed at a value of 2. The exact compositions are given in Table 2.3.

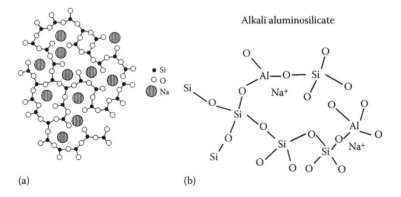

(a) (b)

FIGURE 2.7 Schematic diagram of the NBOs in (a) an alkali silicate and (b) an alkali aluminosilicate.

TABLE 2.2

Typical Aluminosilicate Compositions

Composition	23	33	43	23	DT
Oxide	Wt%	Wt%	Wt%	Wt%	Wt%
SiO_2	66.02	57.64	64.6	61.55	60.7
Al_2O_3	13.62	21.2	13.9	19.76	12.5
B_2O_3		7.27	5.11	3.85	
Na_2O	13.73	12.78	13.75	13.1	12.1
K_2O	1.73	0.73			5.9
MgO	3.95	0.03	2.38	1.44	6.6
CaO	0.45	0.08	0.14	0.04	0.21
BaO					0.08
ZrO_2				0.02	0.98
SnO_2	0.44	0.22	0.08	0.21	
SrO					0.08
$R_2O-Al_2O_3$	6.6	0.7	–0.2	1	8.6

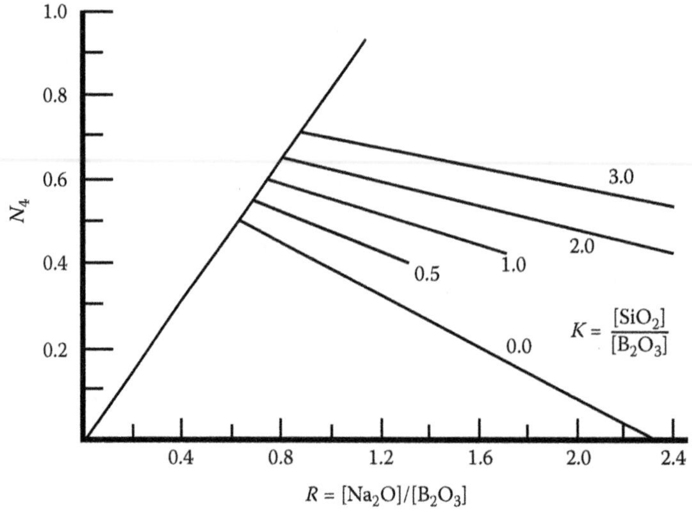

FIGURE 2.8 Schematic representation of the number of 4-coordinated borons as a function of the alkali to boron ratio at a different SiO_2/B_2O_3 ratio (see Figure 5.3).

Experimentally, one finds photosensitivity expresses in the classic Ag photosensitive color developed only in the glasses that contain NBOs. Thus, in both cases for aluminosilicates and borosilicates, the necessary condition to see noble metal photosensitivity is the presence of NBOs. The possible role of NBOs is discussed further in Chapter 7, which discusses solarization (UV-induced color centers).

TABLE 2.3
Alkali Borosilicate Composition Chosen to Test the NBO Relationship to Photosensitivity by Selecting the First Two Compositions without NBOs and the Latter Two with NBOs

	Mole %			
Composition	A	B	C	D
SiO_2	60	55	50	45
B_2O_3	30	27.5	25	22.5
Na_2O	10	17.5	25	32.5
Ag	0.0018	0.0018	0.0018	0.0018
CeO_2	0.014	0.014	0.014	0.014

2.4.2 COPPER-BASED

There is another photosensitive system based on the inclusion of a small amount of Cu (0.6–0.9 wt% CuO) into a glass composition space. An example of the compositional system is listed in Table 2.4.[8,9]

These glasses are clear "as-made" indicating that all of the Cu present is in the Cu^{+1} state. We mentioned in Section 2.2 that the charge transfer transition of Cu^{+1} seemed to satisfy the energy level condition shown in Figure 2.4 to be able to produce a photoelectron with UV light. The induced color is produced in a slightly different manner than in the Ag-based glasses described in the previous section in that the UV exposure is done not at room temperature but at temperatures ranging from 270°C–450°C, with the color appearing after the final heat treatment at 550°C. This is shown in Figure 2.9.

TABLE 2.4
Cu Halide Photosensitive Glass Composition

(Weight%)	CQT
SiO_2	48.2
B_2O_3	20.5
Al_2O_3	8.7
Na_2O	3.4
Li_2O	2.1
K_2O	5.7
CuO	0.4
Cl–	0.54
Br–	0.49
BaO	4.8
SnO_2	0.53
ZrO_2	4.5

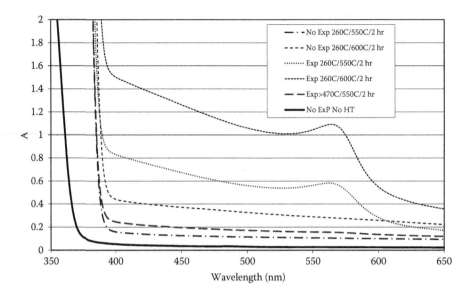

FIGURE 2.9 Spectra of the photosensitively produced color in Cu⁺¹-containing glass as a function of the exposure and thermal treatment noted on the graph.

One can clearly see the surface plasmon resonance of the Cu nanophase at 570 nm developing, which will be described in the next section. The simplest sequence reaction is analogous to Equation 2.1.

$$Cu^+ + hv + heat \rightarrow Cu^{+2} + e \quad Cu^+ + e \rightarrow Cu^0 \quad nCu^0 + heat \rightarrow (Cu^0)_n \quad (2.9)$$

These equations are written analogous to the Ag nanoparticle-based phenomenon with no further justification that the observed formation and optical behavior is similar other than the need for the exposure to be done above room temperature. This may occur because the excited state of the Cu⁺¹ that couples to the conduction band is higher in energy and needs thermal energy to achieve a significant population.

2.4.3 EXAMPLES OF PHOTOSENSITIVE COLOR

In this section examples of patterns made in glass by the photosensitive process producing Ag, Au, and Cu nanoparticles will be shown. Table 2.5 lists the typical exposure sources, the exposure conditions, and the subsequent thermal treatment schedule used to develop the color.

A necessary condition for the wavelength of the exposure source is that it falls within the Ce⁺³ absorption band shown in Figure 2.10a and examples of the colors produced are shown in Figure 2.10b–d. Examples are shown with Ag, which produces the yellow color and with Au and Cu, which produce the reddish hue. We have seen a very slight change with Pt but not with a sharp absorption resonance. The origin of these colors will be discussed in detail in Section 2.5 in addition to the unique optical properties which only certain metals possess that give rise to the color.

TABLE 2.5
Exposure Sources, Exposure Energy, and Thermal Schedules Used for Photosensitive Glasses

Glass	Light Source	Exposure Time	Thermal Treatment
23	1 KW HgXe	10 mW/cm$_2$ – 30 min	650°C/2 hr
02	355 nm	10 Hz/2 W/cm$_2$ – 10 min	480°C/2 hr
Fota-Lite	1 KW HgXe	10 mW/cm$_2$ – 5 min	530°C/30 min + 570°/20 min
Fotoform	1 KW HgXe	10 mW/cm$_2$ – 8 min	560°C/30 min + 600°/30 min
Rainbow	355 nm	10 Hz/2 W/cm$_2$ – 10 min	550°C/2 hr [no ramps]

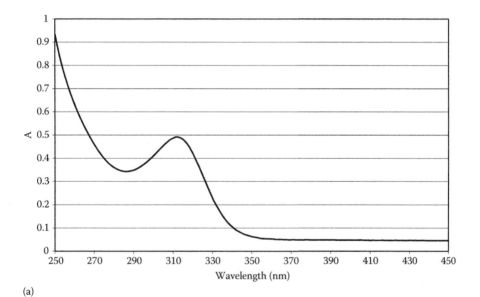

(a)

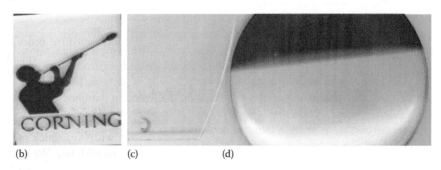

(b)   (c)   (d)

FIGURE 2.10   (a) Absorbance band of a Ce^{+3} ion in photosensitive glass and images of the photosensitively produced colors in photosensitive glasses: (b) gold, (c) Ag, and (d) Cu.

To get some idea of the efficiency of the color development as it pertains to the exposure protocol one may start by writing simple rate equations based on Equation 2.2. For example, one would start with the coupled differential for a number of electrons in the conduction band and those in the traps at any time t.

$$\frac{de}{dt} = kCe^{(3+)}I(\lambda) - \frac{e}{\tau} - \frac{e(T_0 - T)}{\tau_2} \quad \frac{dT}{dt} = e(T_0 - T)/\tau_2 \quad \frac{dAg}{dt} = KIe \quad (2.10)$$

Here, one has the creation term proportional to the product of the intensity the absorption cross section of the Ce^{+3} and the Ce^{+3} concentration, the recombination term, and the term representing the electrons that are trapped. To be complete, a term that represents the possibility of the electron escaping back into the conduction band could be added, which we will neglect. The T_0 term represents the initial number of empty traps. The steady-state value of T is what ultimately counts since it will be these electrons, when thermalized in the subsequent thermal treatment, that can then combine with the Ag^{+1} to form the Ag. The simple result is that the traps will all be eventually filled; that is, $T = T_0$, and at a rate determined by the combination of all rate constants included in Equation 2.8. This demonstrates that the exposure is primarily determined by the total exposure energy joules/cm² and the intensity of the exposure source determines the length of the exposure time. This is consistent with the processes that are described by the rate in Equation 2.10 in that the overall solution for the amount of Ag produced would be of the form $K(T_0)[1-\exp(qIt)]$ with $q(\tau_1,\tau_2)$, where It is the energy of exposure.

2.5   OPTICAL ABSORPTION FROM NANOPARTICLES

All of the photosensitively produced colored glasses covered in the above discussions arise from the absorption of ultimately formed metal nanoparticles from a phenomenon referred to as surface plasmon resonance.[6,10,11] The bulk version is seen in the free-carrier absorption of heavily doped semiconductors. In classical physical terms it is a light-induced electron density wave propagating along the interface between a metal and the dielectric with which it is in contact. It should be mentioned that the plasmon is the quantum mechanical particle version; however, recourse to the quantum development does not add to the physical characterization that we will be dealing with, so we will stay with the classical development but still may use the word *plasmon*. Since the classical point of view is being used to describe the interaction of light with metal, there is recourse to Maxwell's equations and in particular in terms of how one takes into account the way light interacts with a particle. In reality, this is a light-scattering phenomenon where we will be interested in the imaginary part of the scattering cross section that is absorption. The solution to ME equations when the scattering particle is small relative to the wavelength of light so that the electric field of the light can be taken to be constant (quasi-electrostatic approximation) even though the particle is often called the Mie scattering solution.[10] The Mie solution for the absorption coefficient C_{abs} where the extinction coefficient C_{ext} is the sum of the scattering coefficient C_{sc} (real part) and the absorption cross section (imaginary part) is the latter C_{abs}, which is given here in terms of the complex dielectric constant of the metal ϵ and the dielectric constant of the surrounding medium ϵ_m:

$$C_{abs} = \frac{\left(\dfrac{2\pi V \epsilon_m}{\lambda}\right)\epsilon_2}{(\epsilon_1 + 2\epsilon_m)^2 + \epsilon_2^2} \tag{2.11}$$

where $\epsilon = \epsilon_1 + i\epsilon_2$ and in terms of the refractive index since many measurement results are expressed in terms of (n,k):

$$\epsilon_1 = n^2 - k^2 \quad \epsilon_2 = 2nk \tag{2.12}$$

The resonance appears when the real part of the dielectric constant of the metal meets the condition $\epsilon_1 = -2\epsilon_m$; here, ϵ_m is the dielectric constant of the surrounding medium. To get some idea of the frequency dispersion of the dielectric one usually appeals to the Drude model,[11] which essentially is a free electron model (a damped harmonic oscillator approximation) yielding simple expressions for the dielectric functions as a function of frequency and the plasma frequency $(Ne^2/m\epsilon_0)^{1/2}$

$$\epsilon = 1 - \omega_p^2/(\omega^2 - i\gamma\omega) \quad \epsilon_1 = 1 - \frac{\omega_p^2}{\omega^2 + \gamma^2} \quad \epsilon_2 = \frac{\omega_p^2\gamma}{\omega^3 + \omega\gamma^2} \tag{2.13}$$

Here, the physical interpretation of the damping parameter linked inversely to the mean free path (MFP) of the electron is proposed; thus, it is an implicit function of the radius of the metal particle. If the particle radius is smaller than the bulk value of the MFP of the metal then the value of γ in Equation 2.9 would increase inversely proportionally; that is, $\epsilon_2 = \epsilon_{2B} + \text{const}/r$.[12] There is also a quantum mechanical expression for the broadening effect from Kawabata and Kubo[13] that has the same inverse dependence of the particle radius. This size dependence would result in a spectral broadening of the resonance and related slight peak shift. Also to bring Equation 2.9 more in line with real solid materials one brings in an additional term that includes other contributions to the dielectric constant involving electronic interband transitions. This contribution is better understood from the quantum mechanical formulation of the dielectric constant that is cast entirely on the electronic transition between states and their respective transition probabilities.

$$\epsilon = 1 - \Sigma_j \omega_p^2 f_{ij}/\left[\omega(\omega + i\gamma_j) - \omega_{ij}^2\right] \tag{2.14}$$

In this case, the other interband transitions are already embedded in the formula in the term. This interpretation difference aside, in practical terms the mathematical forms of the two formulations are almost identical. As will be seen from the experimental data for ϵ as a function of frequency it is easy to see over what range the simpler form is valid. This stems from the limiting forms of Equation 2.8; namely, when $\omega \gg \gamma$ the above equations reduce to the following:

$$\epsilon_1 = 1 - \omega_p^2/\omega^2 \quad \epsilon_2 = \gamma\omega_p^2/\omega^3 \tag{2.15}$$

This is the free-electron limit so to the extent that the experimental measurements of ϵ follow these forms, one is likely to see the sharp plasma resonance phenomenon.

2.5.1 COMPUTED DIELECTRIC FUNCTIONS

For nanophase metals that can be produced in glass; namely, Ag, Au, Cu, and Pt,[14–16] we will provide the measured optical constants obtained from the literature that are required to see the spectral range in which Equation 2.10 appears to follow the free electron model. Thus, it is appropriate to use Equation 2.7 to predict the spectral position of the surface plasmon resonance that will be used to compare to the experimental data presented in Table 2.6. Using the compiled values of n, k from Palik we can obtain the values of ϵ_1 and ϵ_2 for the metal particles listed above.[17] These are shown in Figure 2.11a,b for Ag, Figure 2.12 for Cu, Figure 2.13 for Au, and Figure 2.14 for Pt.

This data is provided because it is more convenient since we predict the resonance as wavelength rather than frequency and the condition of the resonance when the following is true:

$$-\epsilon_1 = 2\epsilon_m (surrounding\ medium) \tag{2.16}$$

One can see that the free electron behavior; that is, the wavelength (frequency) dependence where the real part of the dielectric constant ϵ_1 as indicated by Equation 2.11, which is observed for all of the metals is some portion of the wavelength (frequency) regime. For Ag it seems to hold quite well over the wavelength region plotted, and a further indication of free electron behavior is that the relationship of ϵ_2 with the wavelength follows the predicted free electron approximation of

TABLE 2.6
Comparison of the Predicted and Observed
Spectral Positions of the Surface Plasmons

Metal	Predicted (nm)	Observed (nm)
Ag	417	420
Au	495	516
Cu	480	570
Pt	370	?

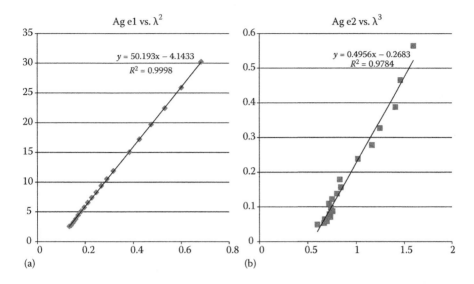

FIGURE 2.11 Plot of real and imaginary parts of the dielectric constant for an Ag fit to the Drude model.

Equation 2.11. For Cu, the region greater than 400 nm appears to confirm free electron behavior; however, the fit of ϵ_2 is poor below 450 nm, making the free electron behavior likely to be confined beyond that wavelength. For Au, the fit to λ^2 form is not as good as it is for either Ag or Cu; however, the Pt data fits quite well for both the real and imaginary parts of the dielectric constant.

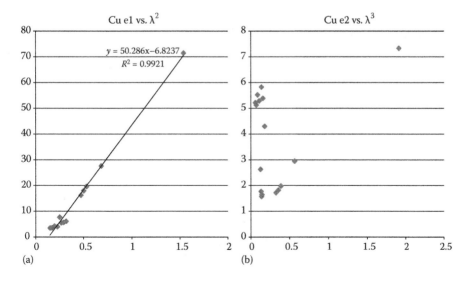

FIGURE 2.12 Real and imaginary parts of the dielectric constants for a Cu fit to the Drude model.

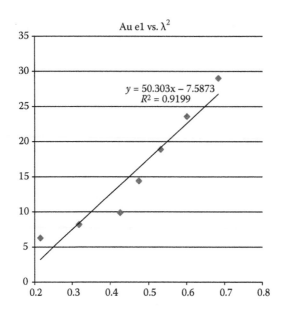

FIGURE 2.13 Real part of the dielectric constants for an Au fit to the Drude model.

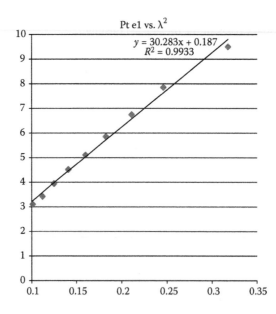

FIGURE 2.14 Real part of the dielectric constants for a Pt fit to the Drude model.

It is only in these wavelengths of the free electron behavior that a well-resolved surface plasmon resonance feature in the absorption spectrum is likely to be seen. The resonance will occur at the wavelength corresponding to Equation 2.12. The sharpness of the resonance even to the point of seeing it all will depend on the relative magnitude of ϵ_2 γ term can be seen from the Lorenztian form of the denominator Equation 2.7.

2.5.2 A COMPARISON OF THE EXPERIMENTAL RESULTS

The measured spectra for the various metal particles are shown in Figure 2.15 through Figure 2.18 for Ag, Cu, Au, and Pt, respectively. The base glass for the Pt was not the same; however, the refractive index that determines ϵ_m was the same. We can compare the predicted position of the resonance from Equation 2.12 using the dielectric data of Figure 2.11 through Figure 2.14 observed from the respective spectra of Figure 2.15 through Figure 2.18.

The agreement between the predicted position of the resonance peaks from the dielectric data and what is observed experimentally from the absorption data is fairly good except for Pt. The fact that one does not see any clear evidence of a resonance is somewhat surprising and it may be that we did not produce enough Pt nanoparticles. The exposure difference of the Pt sample cannot be seen very well in Figure 2.18 because the contrast is poor. Needless to say, the Pt ion is not nearly as soluble in glass as the other three metal ions.

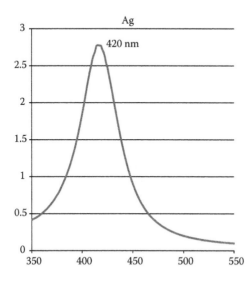

FIGURE 2.15 Measured spectrum of the surface plasmon resonance of an Ag nanoparticle in glass.

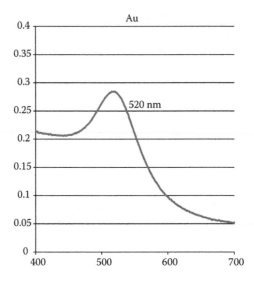

FIGURE 2.16 Measured spectrum of the surface plasmon resonance of an Au nanoparticle in glass.

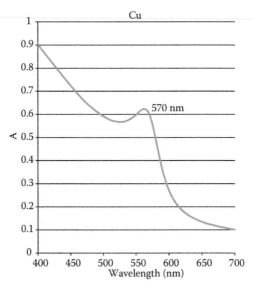

FIGURE 2.17 Measured spectrum of the surface plasmon resonance of a Cu nanoparticle in glass.

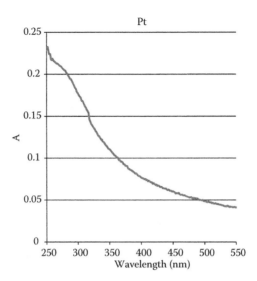

FIGURE 2.18 Measured spectrum of the surface plasmon resonance of a Pt nanoparticle in glass.

REFERENCES

1. S.D. Stookey, Photosensitive gold glass, U.S. Patent Nos. 2,515,937, 1950; 2,515,936, 1950.
2. S.D. Stookey, New photographic medium, *Ind. Eng. Chem.* 41, 856, 1949.
3. S.D. Stookey, Glass by gold silver and copper, *J. Am. Ceram. Soc.* 32, 246, 1949.
4. N.F. Borrelli, in *Microoptics Technology*, Second Edition, Chapter 10, Marcel Dekker, New York, 2005.
5. R.J. Araujo and N. F. Borrelli, *SPIE* 1590, 3424, 1998.
6. R.D. Maurer, Nucleation and growth in a photosensitive glass, *J. Appl. Phys.* 29(1), 1, 1958.
7. N.F. Mott and R.W. Gurney, *Electronic Processes in Ionic Crystals*, Oxford Press, 1948.
8. R.H. Dalton, U.S. Patent No. 2,326,012, August 1947.
9. R.H. Dalton, U.S. Patent No. 2,442,472, June 1947.
10. H.C. Van de Hulst, *Light Scattering by Small Particles*, John Wiley & Sons, New York, 1957.
11. C.F. Bohren and D.R. Huffman, *Absorption and Scattering of Light by Small Particles*, John Wiley & Sons, New York, 1973.
12. U. Kreibig and C.V. Fragstein, *Z. Phys.* 234, 307, 1970.
13. R. Kubo and A. Kawabata, *Annu. Rev. Mater. Sci.* 14, 49, 1984.
14. R.H. Doremus, *J. Chem. Phys.* 40(8), 2389, 1964.
15. U. Kreibig and P. Zacharias, Surface plasmon resonance, *Z. Phys.* 231, 128, 1970.
16. R. Yokaya, *J. Ceram. Soc. Jpn*, 78(8), 39, 1970.
17. E.D. Palik (ed.), *Handbook of Optical Constants of Solids*, Academic Press, San Diego, CA, 1985.

3 Photosensitive Glass-Ceramics

Biscotti: An Italian cookie that is twice baked.

3.1 INTRODUCTION AND BACKGROUND

The subject and content of this chapter continue and significantly add to the material in previous chapter. It is a *continuation* because we start with the formation of the noble metal nanophase in the first step, and it is an *addition* because crystalline nanophases are subsequently produced from the initially formed noble metal particles formed by the method described in Chapter 2. In other words, the photosensitively formed Ag and Au nanoparticles can act as the nuclei for the growth of a variety of crystalline nanophases from the glass matrix. This class of photosensitively produced nanophase materials was initially discovered by S. Donald Stookey while working at the Corning Incorporated (formerly Corning Glass Works) research laboratory in the early 1950s.[1] From then on and even up to the present, new crystalline phases have been produced in this novel and useful way. The advantages stem from the altered chemical, mechanical, and optical properties of the composite glass/crystalline phases with the distinct advantage of being produced essentially by a photolithographic technique; that is, to be able to spatially pattern the regions that are crystalline.

It should be appreciated that all glass-ceramics derive from a glass composition that is in a thermodynamically metastable state produced by the rapid cooling of the glass from a molten state, literally a thousand degrees in a few minutes. Reheating the glass can cause it to crystallize if it is heated above the liquidus temperature and the glass is fluid enough (i.e., the viscosity is sufficiently low at the liquidus temperature). However, it is desirable to control the size of the crystallites, and therefore it is often necessary to provide nucleating agents. This means that it is not necessary to reach as high a temperature, thereby exercising more control of the rate of crystallization. Nucleating agents are constituents added to the glass composition to provide some type of compositional fluctuation that influences the crystalline formation. Nucleating agents can be Ti or Zr or phosphate.[2,3] The relevant point here is that noble metal nanoparticle can provide this function as well. It is beyond the scope of this book to describe what the physical and chemical mechanism(s) are or the various nucleation processes for various phases but it is certainly interesting to speculate exactly how the nucleating process works. In the case of TiO_2 additions, it is speculated that it may form some sort of localized titanate immiscible phase that in turn provides a place where a major crystalline phase can begin to segregate from the glass. From a thermodynamic point of view, one could view the phenomenon as a lowering of the free energy barrier to crystallization by the composition changes in a local area. In other words, any local composition fluctuation alters the free energy and hence produces diffusion, further altering the composition. In all cases we will deal with the phase that is to be produced in a controlled manner which is invariably the phase that would be produced by heating the glass to a higher temperature but with no control of the amount or grain size. In other words, glass-ceramics is really just another word for controlled crystallization. A much more complete discussion of the physics of nucleation is provided in Holand and Beall's book on glass-ceramics technology.[4]

3.2   NOBLE METAL NUCLEATION

In some glasses, certain noble metal particles can nucleate certain phases and even more surprising is that these phases can act as nuclei for additional phases to form the kind of twice-baked situation alluded to at the beginning of the chapter.

The extension of this process to have the Ag nanoparticle act as a nucleation center for a nanocrystalline phase derived from the glass has been known and reported in both the open and patent literature.[5,6] Examples from both will be covered here.

Schematically, one adds another reaction equation, shown below as Equation 3.4, to the simplified version of the schema given in Chapter 2, which is repeated here for convenience:

$$Ce^{+3} + hv \rightarrow e + Ce^{+3+} \ (250\text{–}350 \ nm) \tag{3.1}$$

$$Ag^+ + e + heat \rightarrow Ag_0 \tag{3.2}$$

$$n(Ag_0) + heat \rightarrow (Ag_0)_n \tag{3.3}$$

$$(Ag_0)_n + X + Y + heat \rightarrow XY \tag{3.4}$$

Here, XY represents the nanophase material produced by the glass constituents X and Y nucleated by the Ag nanoparticle. In reality Equations 3.3 and 3.4 occur together as a consequence of the thermal treatment. The chemistry of the actual nucleation process is not known; that is, by what reaction(s) actually occur aided and abetted by the Ag nanoparticle. Clearly, this involves a heterogeneous nucleation mechanism, which has been discussed in some detail in Holand and Beall[4] in the case of a metal agent. It is described in terms of the interplay of the various interfacial energies, metal/liquid, crystal/liquid, and metal/solid, as well as the contact angle. However, there is one study that sheds some light (forgive the pun) on the initial stage of the process. Borrelli et al.[5] used the shift in the spectral position of the silver plasmon resonance that is a function of the complex dielectric constant of the Ag and the dielectric constant of the medium ϵ_m in which the Ag is embedded (see Section 2.5) to track the change in composition surrounding the Ag nuclei as a function of the thermal treatment. We copy the expression for the absorption here for convenience of the reader. Note how the spectral position of the resonance absorption depends on the condition when $\epsilon_1 = -2\epsilon_m$.

$$C_{abs} = \left[\frac{2\pi V \epsilon_m}{\lambda}\right]\left[\frac{\epsilon_2}{(\epsilon_1 + 2\epsilon_m)^2 + \epsilon_2^2}\right] \qquad (3.5)$$

A glass was chosen to study the halide crystal growth for a typical photosensitive glass composition where the halide in this case was Br.

Figure 3.1 shows a typical transmission spectra after a 560°C heat treatment for the glass without bromine and with bromine, respectively. For the bromine-free glass, the absorption peak was seen to shift only slightly in wavelength as a function of the temperature of the thermal treatment. This indicated that only a small effect due to the change in the silver speck size occurred. The variation of the absorption peak was used to estimate the silver speck size as a function of the thermal treatment by the use of the above expression.[5] The size effect on the absorption is due to the size dependence of ϵ_2 expressed by $\epsilon_2 = \epsilon_{2B} + C/r$ as discussed in Section 2.5. The absorption peak shift was used rather than the change in the absorption band half-width because the latter was significantly altered by the absorption edge owing to the cerium contained in the glass. The inset in Figure 3.2 shows the computed radius versus the heat-treatment temperature. Using this computed variation of the silver particle size with heat-treating temperature, the absorption data for the bromine-containing glass was analyzed to determine the value of the refractive index of the surrounding medium required to match the observed wavelength of maximum absorption at each temperature. The result of the calculation is shown in Figure 3.2.

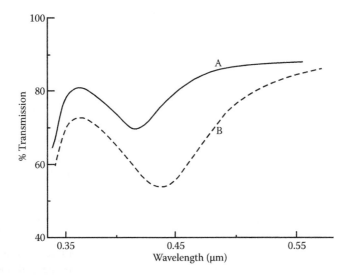

FIGURE 3.1 Surface plasmon absorption of an Ag nanoparticle in a glass with and without Br to show that in the glass containing Br, the refractive index surrounding the Ag particle is being increased by Br enrichment, thus shifting the peak position according to Equation 3.5. (From N.F. Borrelli, J.B. Chodak, D.A. Nolan, and T.P. Seward, *J. Opt. Soc. Am.* 69(11), 1514, 1979.[5])

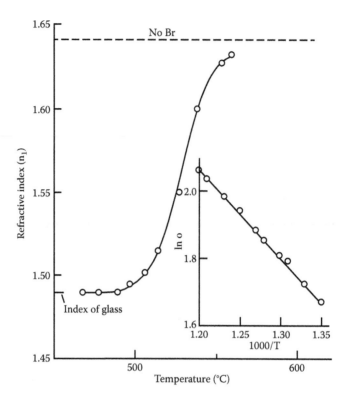

FIGURE 3.2 Computed refractive index of the region surrounding the Ag speck from the shift in the peak of the Ag surface plasmon absorption to indicate the Br enrichment. Inset shows the estimated change in the Ag particle size. (From N.F. Borrelli, J.B. Chodak, D.A. Nolan, and T.P. Seward, *J. Opt. Soc. Am.* 69(11), 1514, 1979.[5])

As can be seen, the index of the surrounding medium rises sharply at slightly above 500°C and assumes a value essentially that of NaBr at about 560°C. This can be explained by the mechanism that, at just above 500°C, a NaBr enrichment of the region about the silver speck begins to occur, and with a higher heat-treating temperature, leads ultimately to a pure sodium bromide phase. This observation indicates that the silver nucleation of a sodium halide phase occurs initially by enriching the surrounding region with Br. This is likely the effect in all cases where noble metal nucleation is seen that is an enrichment of the ions that ultimately will form a crystal. Nucleation is a good example of the statistical thermodynamic fluctuation phenomenon, where any perturbation in the local concentration of constituents in this case caused by the formation of the Ag nanocrystal alters the chemical potential that then drives diffusion, further altering the local environment. It is the same argument irrespective of what crystal is ultimately formed. When the nucleating agent is yet another nanocrystalline phase, as we will be discussed in Section 3.6, it becomes more difficult to define a specific mechanism where each is unique to the crystal that will form.

We see this in another phenomenon that will be described in Section 3.6, where the process can continue through a second stage. Here, the initial photosensitive nanocrystals produced by the mechanism described above[8] further nucleates an entirely different crystalline phase from the glass upon heating to a higher temperature.

This suggests yet another reaction to be added to the above set where the new phase (XYZ), which is compositional constituents of the glass, is produced by the nucleation of (AB), which was photosensitively formed through Equations 3.1 to 3.4.

$$AB + X + Y + Z + heat \rightarrow XYZ \tag{3.6}$$

In the actual process, we will see that steps 3–5 all occur together although the temperature is often higher than required for the steps expressed by Equations 3.3, 3.4, and 3.6 to proceed.

We start with two examples where the Ag nanoparticles nucleate a nanocrystalline phase. The first will be a description of how a NaF phase is developed and the second will be where a Li-metasilicate is formed. For both examples, some of the unique properties will also be mentioned.

We then move to the next level of photosensitively produced crystalline nanophase by covering the nucleation of more complex phases from the glass constituents from the photonucleated NaF and/or LiF phases. This is an entirely new twist on the photosensitive process—having the initially produced nanocrystals nucleate yet another phase, the alkali-aluminate phase, which would have been produced thermally at a higher temperature. This will be the first example of a truly photosensitively produced glass-ceramic defined as having more than 50% crystal. This opens up the study of a new phenomenon of three separate nucleation processes (Ag, Na/LiF, Na/Li-aluminate), which all occur in one glass sequentially in time.

3.3   FOTA-LITE

Fota-Lite is a Corning Incorporated trademark for a photosensitive NaF-based opal glass, the composition of which is given in Table 3.1.[6–8]

TABLE 3.1
Composition of Fota-Lite with Descriptions of the Function of the Different Components

Glass Component	Wt%	Function
SiO_2	69.0	
Na_2O	15.8	
ZnO	4.8	Glass matrix
Al_2O_3	6.8	
F^-	2.3	Crystal constituents (with Na and Ag)
Br^-	1.0	
Ag^+	0.01	Sensitizer and colorant
CeO_2	0.05	Optical sensitizer
Sb_2O_3	0.20	Thermal sensitizers, redox agents, refining agents
SnO	0.05	

The key element is the fluorine component. At high temperature the fluorine in the glass is likely incorporated into the silica network as a Si-F species. There is a limited amount of fluorine that can be maintained: <3% by weight mainly because the fluorine is so volatile at the melting temperature. When the glass is reheated >600°C, a NaF opal phase results, indicating that the fluorine as an Si-F bond is thermodynamically metastable and would prefer to react with a mobile alkali ion, in this case sodium, to form a white opal glass. The amount of the NaF formed can be estimated from the measured fluorine content and would be estimated to be a few volume percent. As discussed in Chapter 2, with the addition of Ce/Ag to this base glass composition, which has a high NBO concentration, the glass can be made photosensitive (i.e., produce an Ag nanoparticle by exposure to light and a thermal treatment). The resulting Ag nanoparticle acts as a nucleus to produce the NaF nano-phase at a lower temperature (500°C) than the spontaneous formation of a NaF phase (>600°C). The result pertains to a glass where only the region that was exposed to UV light after thermal treatment will produce a white opal phase. The exposure conditions and thermal development schedule are provided in Table 3.2 and Figure 3.3.

TABLE 3.2

Exposure Conditions for the Development of the NaF Phase

Lamp: 1 kW HgXe
Intensity: 10 mw/cm²
Wavelength: 30 nm
Time: 1–5 minutes
Total fluence: ~6 J/cm²

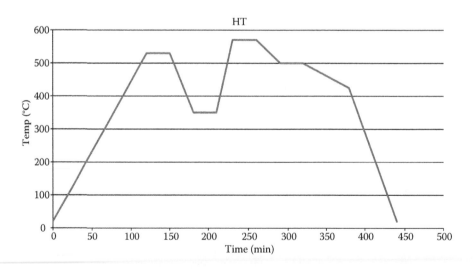

FIGURE 3.3 Typical thermal development schedule for the development of the NaF phase in Fota-Lite.

The glass can be spatially patterned using a suitable light blocking mask that produces a pattern, as shown in Figure 3.4.

The TEM image and XRD of the NaF particles are shown in Figure 3.5. Figure 3.5a shows the unexposed state that is heated and also shows some evidence of the 10–20-nm NaF particles (the dark spots). Figure 3.5b shows the exposed and heated state and although the dark spots are still evident what appears is clear evidence of a true phase separation superimposed on the NaF nanoparticles. This accounts for the high degree of scattering leading to the opaque state. As we will see in Section 3.4 and in Section 4.2 different optical properties are evident in the exposed state albeit with a different exposure and thermal development schedule. This does suggest that the opacity of the Fota-Lite glass is not solely due to simply the result of the formation of sparse NaF nanoparticles but what appears to be a cloud of particles brought about by the specific exposure/thermal treatment. This distinction in appearance was not

FIGURE 3.4   Pattern formed in Fota-Lite by exposure through a photomask followed by the thermal development shown in Figure 3.3.

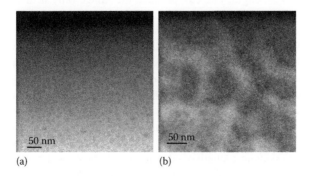

(a)   (b)

FIGURE 3.5   Thermally developed TEM images of Fota-Lite: (a) unexposed and (b) exposed region.   (*Continued*)

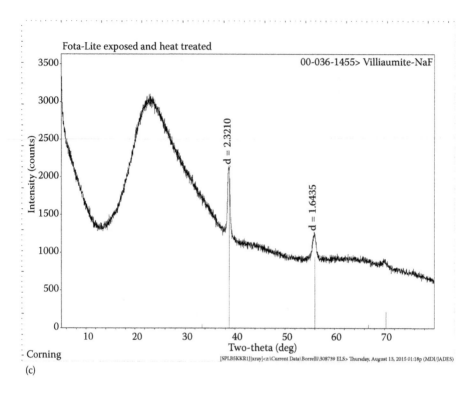

FIGURE 3.5 (CONTINUED) Thermally developed TEM images of Fota-Lite: (c) XRD trace of an exposed and developed phase.

previously appreciated, as we will see in the discussion of the other optical effects that can be produced in Fota-Lite glass. Figure 3.5c shows the XRD trace.

The resolution of any image formation can be quite good since the NaF nanoparticles are <20 nm, although to utilize it the depth of the exposure must be limited. Because the exposure depth of the image can be controlled by the absorption depth at the excitation wavelength, one can produce the image only near the surface, as shown in in the cross section image in Figure 3.6.

The exposure in the first case was done with a 248-nm excimer laser and the second with a 193-nm excimer. Chapter 5 discusses the photorefractive phenomenon where we will show how this glass can be used as a holographic medium with a slight modification in composition to limit the NaF particle size to maintain transparency.

Another consequence of patterning is the ability to make a gradient scattering element, as shown in Figure 3.7.

This is accomplished by using a photomask with an array of small openings spaced to have the fill factor varied along the length. An application[9] of this is a way to produce uniform illumination along a length of edge-illuminated bar, as shown in Figure 3.8.

One can further extend this structure to an edge-illuminated array of LEDs where the phosphor layer is placed on top of the patterned Fota-Lite bar, as shown in Figure 3.9.

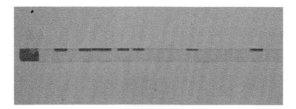

FIGURE 3.6 Deep UV laser exposures to limit the opal region to the surface.

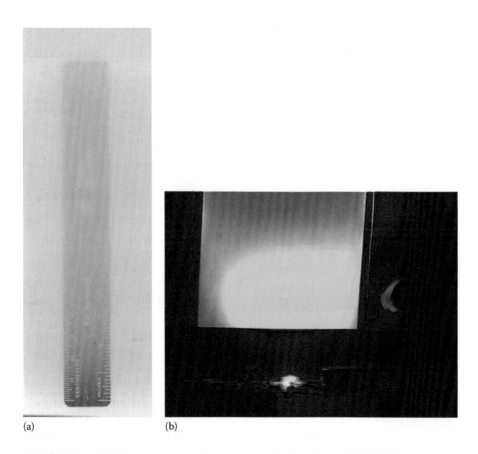

(a)   (b)

FIGURE 3.7 (a) Patterned exposure to produce gradient scattering, (b) LED illumination of (a) to produce more uniform scattering.

3.4 POLYCHROMATIC GLASS

Perhaps one of the most fascinating phenomena taken from all of the varieties of effects described in this book is the polychromatic effect.[10] Not only fascinating in its ability to make a rainbow of colors but the multifaceted way (literally as well as

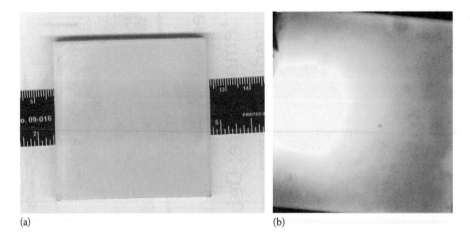

(a)   (b)

FIGURE 3.8   LED illumination on the edge of a gradient scatterer that is (a) phosphor-coated and (b) LED-illuminated. (From N.F. Borrelli, N. Lonnroth et al. U.S. Patent No. 9,011,720.[9])

(a)   (b)

FIGURE 3.9   Photographs of the developed colors as indicated in Table 3.4.

figuratively, as we will see) the effect develops over the exposure time and the thermal processing.

One starts with essentially the same composition given above for Fota-Lite except for the addition of a somewhat stronger reducing action produced by the addition of SnO, as shown in Table 3.1. The thermal and physical properties are shown in Table 3.3.

The reducing action stems from the ability of Sn^{+2} to give up electrons by going to Sn^{+4} during the thermal treatment, which will be important in the reduction of the Ag later in the coloring process. The Ag still plays the same role as it does in all the photosensitive glasses in forming the initial Ag nanoparticle that then nucleates the NaBrF phase in this case. What is very different is its transparency in spite of

TABLE 3.3
Thermal and Mechanical Properties of Polychromatic Glass

Softening point	664
Annealing point	484
Strain point	444
Thermal expansion ($\times 10^{-7}$)	83
Density	2.482
Other properties	Liquidus temperature = A-831°C I-834°C P-837°C; liquid viscosity = 300,000
Composition	Alkali-zinc aluminasilicate

little compositional difference is that the polychromatic version of the NaBrF-based photosensitive glass is its transparency relative to the dense to transparent opal of the Fota-Lite. This is clearly related to the size of the nanocrystals and some of this is likely due to the somewhat lower temperature of the thermal development, as we will see in the thermal treatments discussed next.

3.4.1 PROCESSES

3.4.1.1 First Exposure and Thermal Treatment

A 1-kW Hg/Xe UV source is used for all of the Fota-Lite exposures provided in Section 3.3 (Table 3.2), but in the case of polychromatic glasses, the exposure times are of a much shorter duration. In addition, the initial thermal development was done at 10–20 degree lower temperature.

3.4.1.2 Second Exposure and Thermal Treatment

This important coloring step is where the glass from the first treatment is then exposed with UV again but now the exposure is done at >300°C and from tens of minutes to hours. The longer the second exposure times at the elevated temperature, the deeper the color. Figure 3.9a,b shows examples of the colors that can be developed by the treatments listed in Table 3.4.

The measured transmission as a function of the first exposure time as marked on the curve after development of the exposure and development is shown in Figure 3.10.

It is important to note the progression of the transmission minima (absorption maxima) moves to longer wavelength the longer the first exposure time. This behavior will be explained in Section 3.5 and will be related to the aspect ratio noted in Table 3.4.

3.4.2 MICROSTRUCTURES AND MECHANISMS

The morphology of the nanocrystals obtained from the TEM responsible for the colors is shown in Figure 3.11.

TABLE 3.4

Developed Colors as a Function of Second Exposure Times with Estimates of Particle Size and Model Prediction of the Correlation to the Aspect Ratio

Color	First Exposure Time[a] (sec)	Particle Length (Å)	Particle Width (Maximum) (Å)	Eccentricity (l/w)	Absorption Maxima Observed (Å)	Theoretical, Assuming Ellipsoids with Axes l,w (Å)
Pale green	10	360	60	6.0	770	995
Turquoise	20	210	45	4.7	640	845
Blue	30	160	40	4.0	580	770
Purple	45	120	35	3.4	540	700
Magenta	60	100	35	2.9	520	635
Red	75	75	30	2.5	500	590
Red-orange	105	60	30	2.0	480	535
Orange	135	50	30	1.7	460	495
Deep yellow	195	40	30	1.3	440	455
Yellow	∞	30	30	1.0	415	415

Source: S.D. Stookey, G.H. Beall, and J.E. Pierson, *J. Appl. Phys.* 49, 5114, 1978.[10]

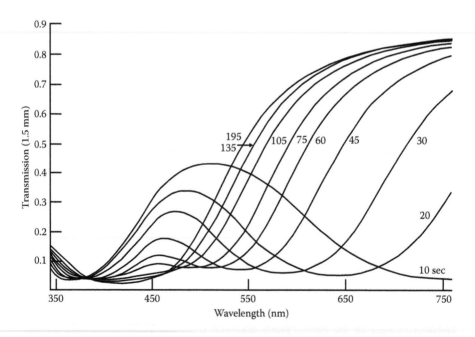

FIGURE 3.10 Transmission spectra of the various colors shown in Figure 3.9. (From S.D. Stookey, G.H. Beall, and J.E. Pierson, *J. Appl. Phys.* 49, 5114, 1978.[10])

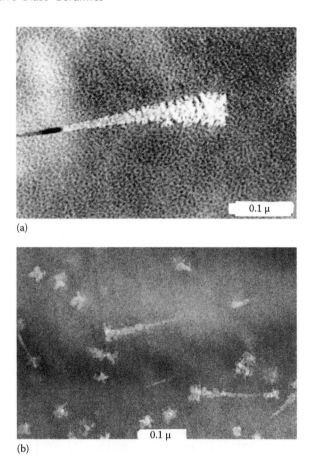

(a)

(b)

FIGURE 3.11 Two representative TEM images (a) and (b) of the NaF nanoparticles. (From S.D. Stookey, G.H. Beall, and J.E. Pierson, *J. Appl. Phys.* 49, 5114, 1978.[10])

Surprisingly, after the initial exposure and thermal treatment (the first step in Section 3.4.1), instead of seeing the expected cube-shaped NaF crystals one sees a number of them with a pyramidal growth from only one face of the cube. It is proposed that the presence of these tails eventually produces the color in the second step. The proposed model is shown in Figure 3.12 where the compositional makeup of the tail is considered to be continually changing along the length and Ag-rich at the tip. NaF, NaBr, and AgBr are all body-centered crystals and could easily form solid solutions.

It is important to bear in mind that there is excess Ag in the glass over and above what is used in the formation of the Ag nucleus. The silver-rich tip is clearly anisotropically shaped and to the extent that it can be approximated as an ellipsoid of revolution (i.e., as having two surface plasmon resonances, one along the length and the other in the short direction), one can make a comparison of the spectra position of the measured absorption. This is shown in Table 3.4. The reader is referred to a fuller discussion in Stookey et al.[10] and calculations using

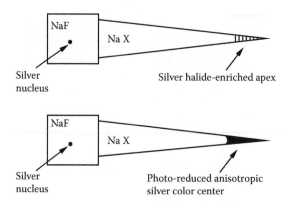

FIGURE 3.12 Schematic drawing approximation of the shape of the pyramidal structure showing the reduced Ag tip, the length of which will be used in the modeling to account for the color. (From N.F. Borrelli, J.B. Chodak, D.A. Nolan, and T.P. Seward, *J. Opt. Soc. Am.* 69(11), 1514, 1979.[5])

different approximate shapes in this reference, as shown in Figure 3.13. Suffice it to say, that the simplest model was chosen; Figure 3.13 was used to calculate the position of the surface plasmon absorption using the following equation where V is the volume of the particle, n_0 is the refractive index of the glass matrix, λ is the wavelength in free space, and L is the electric depolarization factor appropriate for the particle geometry and orientation with respect to the applied field.

$$C_{abs} = \left(2\pi V n_0^3 / L^2 \lambda\right) \cdot \left(\varepsilon_2 \big/ \left\{\left[\varepsilon_1 + n_0^2(1/L - 1)\right]^2 + \varepsilon_2^2\right\}\right) \tag{3.6}$$

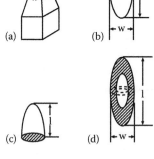

FIGURE 3.13 Possible model structures to approximate the pyramidal tip for the calculation of the surface plasmon resonant absorption: (a) pyramidal tip, (b) ellipsoid, (c) half-ellipsoid, and (d) elliptical shell. (From N.F. Borrelli, J.B. Chodak, D.A. Nolan, and T.P. Seward, *J. Opt. Soc. Am.* 69(11), 1514, 1979.[5])

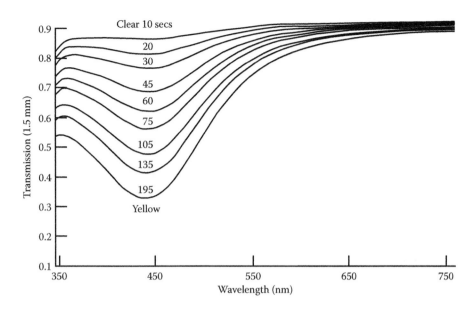

FIGURE 3.14 Strength of the Ag surface plasmon absorption as a function of time of the first exposure. (From S.D. Stookey, G.H. Beall, and J.E. Pierson, *J. Appl. Phys.* 49, 5114, 1978.[10])

What is not well understood is the role of the initial exposure as the determining step leading to a particular color. If we look at Figure 3.10, we see that the transmission minimum (absorption maximum) is at the longest wavelength for the shortest initial exposure and then moves to the blue as the first exposure is increased. Looking at Table 3.4, it appears that the shortest exposure produces the longest tip. It can be argued that the longer first exposure frustrates the growth of the pyramidal tip for a reason that is not clear. The role of the second exposure is to reduce the silver-rich tip to silver metal; the longer the second exposure, the deeper the color. If we look at the absorption after the first step we see an increase of the Ag plasmon resonance alone, which more than likely indicates the growth of more Ag nuclei and thus more nanocrystals (see Figure 3.14). A possible explanation for this phenomenon is that the longer exposure produces more nuclei at perhaps the expense of longer pyramidal growth.

3.4.3 COLORING FOTA-LIGHT

Another example of the variety of optical effects we can observe in NaF-based photosensitive glass is the ability to produce color, which is produced after the photosensitive NaF phase has been made (see Figure 3.15). As in the polychromatic case described above, the sample is exposed while at a temperature of 200°C–400°C for an hour or at which point colors appear, tinting the opal phase. The complete exposure thermal treatment protocol is shown in Table 3.5 and the spectral data is shown in Figure 3.16.

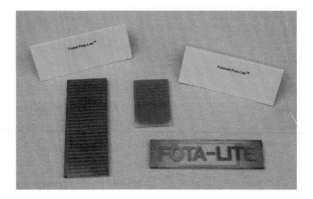

FIGURE 3.15   Photo of the colors produced in Fota-Lite.

TABLE 3.5
Exposure Times and Thermal Treatments for Colored Fota-Lite

Sample	Step 1 Xe Lamp Exposure	Step 2 Heat Treatment	Step 3 Xe Lamp Exposure
1	30 seconds		
2	45 seconds	540°C/40 min, cool,	Flood expose while on
3	60 seconds	560°C/20 min	hot plate setting
4	90 seconds		3.5—several hours

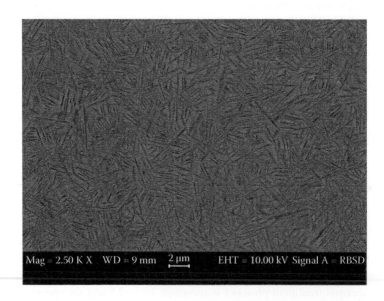

FIGURE 3.16   SEM showing the structure of exposed and thermally developed Fotoform.

The colors are weaker than in the polychromatic version. The hue depends on how long the initial exposure to develop the NaF phase is. The mechanism is the same as the one provided in Section 3.4 on polychromatic glass, since it is produced by the same effect.

3.5   FOTOFORM

Fotoform is another of the noble metal nucleated phases but one that is quite different from Foto-Lite in two important and distinctive ways. The first is that the phase developed from the glass constituents of the base composition is not from any special additional components such as F and Br in the polychromatic case. The second difference is that it significantly alters the mechanical and chemical properties of the parent glass. In other words, it transforms the glass into a glass-ceramic with unique properties. Strictly speaking, it still does not legally qualify as a glass-ceramic because it is only ~20% crystalline; nonetheless, it shares many of the same characteristics. This is another of Stookey's[6–8] discoveries in the late 1950s, which ironically was a temperature overshoot accident in his study of Fotoform that produced the first Li-aluminosilicate glass-ceramic. *Even his mistakes turned into significant inventions!*

Fotoform is essentially a mixed alkali aluminosilicate. It is unstable upon heating above 650°C and produces a Li-metasilicate phase whereby heating to a higher temperature ~800°C converts it to a Li-disilicate phase which also has additional different and interesting properties that will be covered in Section 3.5.3.

The base composition of Fotoform is shown in Table 3.6 and a listing of its thermal and mechanical properties after exposure and thermal treatment to produce the glass-ceramic material is shown in Table 3.7.

So once again, we have an unstable glass composition that when heated above a certain temperature crystallizes spontaneously. However, by adding the photosensitizing agents of Ce and Ag and then exposing the glass to UV light and thermally developing the Ag nuclei, we can now produce the Li-metasilicate phase at a temperature below 600°C. The exposure and thermal treatment are listed in Table 3.8.

TABLE 3.6

Composition of Fotoform

Fotoform/Fotoceram (Corning Code 8603)	Wt%
SiO_2	79.6
Al_2O_3	4.0
Li_2O	9.3
K_2O	4.1
Na_2O	1.6
Ag	0.11
Au	0.001
CeO_2	0.014
SnO_2	0.003
Sb_2O_3	0.4

TABLE 3.7
Thermal and Mechanical Properties of Fotoform

Properties	Fotoform Glass	Fotoform Opal Glass-Ceramic
Mechanical		
Density/g/cm^3	2.365	2.380
Young's modulus/		
·10^6 psi	11.15	12.00
GPa	77	83
Modulus of rupture, abraded/		
psi	8690	12,100
MPa	60	83
Poisson's ratio	0.22	0.21
Hardness (Knoop)/KHN$_{100}$		
Wilson Turkon	450	500
Thermal		
Coefficient of thermal		
Expansion (25°C–300°C)/K^{-1}	$8.4 \cdot 10^{-6}$	$8.9 \cdot 10^{-6}$
Thermal conductivity/W/(m·K)		
25°C	0.75	1.5
200°C	1.1	1.5
Maximum safe processing	450	550
temperature/°C		
Specific heat/J/(g·K)		
25°C	0.88	0.88
200°C	1.2	1.2
Electrical		
log$_{10}$ volume resistivity/Ω cm		
250°C	6.27	8.81
350°C	4.90	7.23
Dissipation factor, at 100 KHz		
21°C	0.008	0.004
150°C	0.050	0.014
Dielectric constant, at 100 KHz		
21°C	7.62	5.73
150°C	9.06	6.14
Loss factor, at 100 KHz		
21°C	0.061	0.023
Dielectric strength/volts/mil to 10 mil		
sample thickness, DC under oil		
25°C	4500	4000

TABLE 3.8

Exposure and Thermal Treatments Used for Fotoform

Source	Energy Delivered	Thermal Treatment
HgXe 1-kW lamp	~10 mW/cm²/(4–8 min)	540°C/30 min, 600°C/30 min
254-nm germicidal lamp	~5 mW/cm²/(2–7 min)	540°C/30 min, 600°C/30 min
Tripled YAG	200 kHz (400 μJ)/355-nm translation 50 mm/s	540°C/30 min, 600°C/30 min
Quadrupled YAG	1 kHz/266 nm/2W-translation 50 mm/s	540°C/30 min, 600°C/30 min
Ti-sapphire femtosecond laser	(~2 μJ) 40 fs/20 kHz-/800 nm-translation 20 μm/s	540°C/30 min, 600°C/30 min
172-nm excimer lamp	172 nm (5–60 min)	540°C/30 min, 600°C/30 min
KrF excimer laser	193 nm (2J) 1–20 shots	540°C/30 min, 600°C/30 min
ArF excimer laser	248 nm (2J) 1–20 shots	540°C/30 min, 600°C/30 min

The following section covers the different unique aspects of the material and the applications thereof:

1. Chemically machinable due to the large difference in HF solubility between the glass and the crystallized state
2. Controlled CTE composite material produced by a combination of exposure and thermal schedules
3. Microoptics lenses and lens arrays utilizing large density difference between crystallized (exposed) regions and the glass (unexposed) regions

3.5.1 CHEMICAL MACHINABILITY

Chemical machinability is the term used to describe one of the distinctive properties of the Fotoform material (note that we will use FF as an abbreviation from here on for convenience) in that the photosensitively produced Li-metasilicate phase (Li_2SiO_3) is roughly 20 times more soluble in diluted HF than the unexposed glass. The differential etching is facilitated by the overlapping and contiguous morphology of the dendritic crystallites, as indicated in Figure 3.16.

It is the property that the grains form a contiguous overlapping pattern with little or no intervening glass that makes the etch rate so much greater than the glass. The etch rate of the developed region is of the order of 13 μm/minute depending somewhat on the specific exposure/thermal treatment. This photosensitive property allows one to then optically pattern the glass using standard photolithographic techniques to etch visas or holes completely through a 1–2 mm thick piece of glass. Some examples of different size hole patterns are shown in Figure 3.17.

The resolution of any pattern is limited by the size of the metasilicate grain, which is about 50 μm. A typical exposure using a 1-kW HgXe lamp with a 360-nm intensity of 10 mW/cm² is 8 minutes so the energy required is of the order of 5 J/cm² to develop the dense crystalline phase. By using a shorter wavelength, such as a

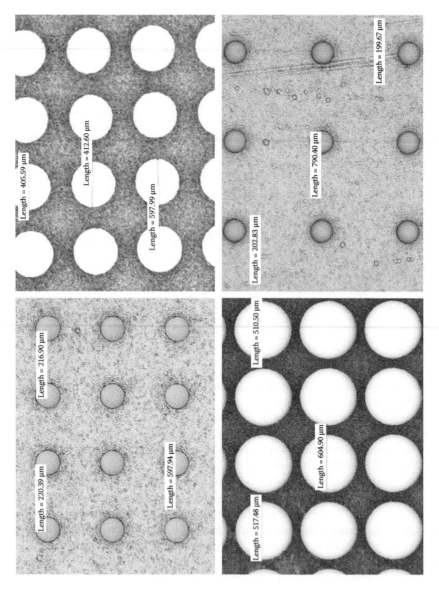

FIGURE 3.17 Chemically etched hole pattern in exposed and thermally developed Fotoform.

248-nm excimer laser ~1 W/cm² ~ 0.1 J/cm² per pulse, we can limit the depth, as shown in Figure 3.18.

A focused 800-nm fs-laser can also be used, which produces sufficiently high intensity that multiphoton processes will occur.[11] This means that the excitation of the photosensitizer Ce^{+3} could be operative even with 800-nm light. In fact, it has been observed that Ce^{+3} is not necessary that is a photoelectron can be produced directly from the valence and of the glass by a multiphoton process. The multiphoton process would be restricted to the high intensity (focused) portion of the beam. This can be estimated from the confocal distance, which is approximately $\lambda/(NA)^2$. Using an 800-nm light one can initiate the photoreaction and the subsequent crystallization anywhere inside the glass because the glass is transparent at 800 nm. Figure 3.19 shows the exposure pattern. A distinct advantage of a multiphoton process is that the

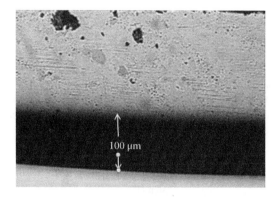

FIGURE 3.18 A 248-nm excimer laser exposure to produce shallow depth of penetration.

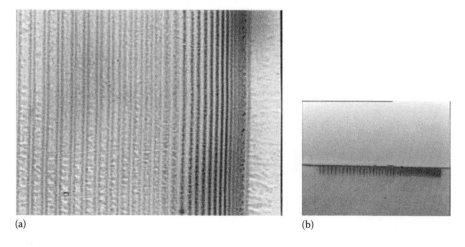

(a)   (b)

FIGURE 3.19 An 800-nm femtosecond exposure of (a) a chirped grating, and (b) a micrograph of the depth of the grating.

beam is capable of focusing just below the surface. The cross-section view shows the depth of the exposure. For the 100-nJ/cm^2 exposure the depth was 100 μm and for the 200-nJ/cm^2 exposure it was 150 μm.

3.5.2 SMILE™ LENS ARRAYS

Another interesting application of Fotoform is that it provides a way to make micro-lens arrays utilizing the effect.[11,12] When the metasilicate phase is produced by exposure and thermal treatment, that region is more dense by ~15% than the nonexposed glass. This density difference can be used to produce surface features since the temperature used to develop the crystalline phase (600°C) means that the unexposed glass is soft and can therefore flow under the influence of the densification of the surrounding material. The net effect is that the exposed area can squeeze the glassy nonexposed region above the surface analogous to squeezing toothpaste out of a tube. If one then uses a photomask where the light is blocked in circular regions, the soft glass will be squeezed out by denser surrounding crystalline material to above the surface and then because of surface tension the soft glass will form a spherical surface. The process is shown in Figure 3.20. An SEM photo of the formed lenses in shown in Figure 3.21.

The applications for the lens arrays made by this essentially photolithographic technique are seen in scanner lens bars and optical fiber interconnects, among others.[13]

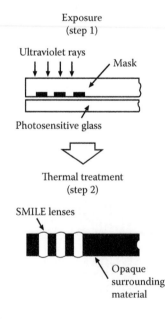

FIGURE 3.20 Schematic representation of the way lenses are formed in a Fotoform material by exposure and thermal development. (From N.F. Borrelli, D.L. Morse, R.H. Bellman, and W.L. Morgan, *J. Appl. Phys.* 24, 2520, 1985.[11])

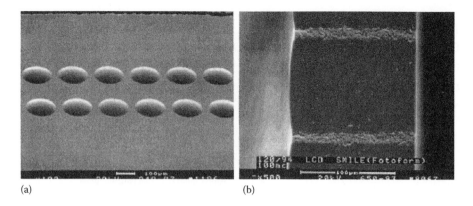

(a) (b)

FIGURE 3.21 (a) SEM photographs of the lenses formed by the process shown in Figure 3.20, and (b) an SEM picture of a cross section of a single lens. (From N.F. Borrelli, D.L. Morse, R.H. Bellman, and W.L. Morgan, *J. Appl. Phys.* 24, 2520, 1985.[11])

3.5.3 CONTROLLED CTE FOTOFORM

The composition of Fotoform can be modified in such a way as to have other phases produced thermally by heating to a higher temperature (>800°C), as shown in Table 3.9. In particular, the addition of alumina for silica allows for a number of lithium alumino-silicate (LAS) phases to be possibly formed at the higher temperatures (see Holand and Beall's book[4] for the Li_2O-Al_2O_3-SiO_2 phase diagram, reproduced here as Table 3.9).

TABLE 3.9
Phase Diagram of the Li-Aluminosilicate System

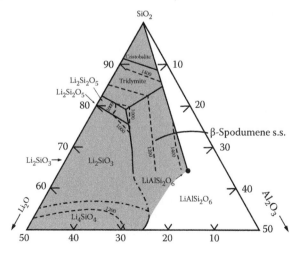

Source: W. Holand and G.H. Beall, *Glass-Ceramic Technology*, Second Edition, John Wiley & Sons, Hoboken, NJ, 2012.[4]

TABLE 3.10

Compilation of Exposure and Thermal Development Protocols Used for Producing a Material with a Variety of Composite CTE Materials

Condition/Treatment	CTE (ppm)	Phase Assembly
No exposure, 900°C/2h	3	Beta-spodumene, Li-disilicate, glass
No exposure, 875°C/2h	4	Less spodumene, Li-disilicate, glass
Exposed, 850°C/2h	5	Spodumene, Li-disilicate, Li-metasilicate, glass
Exposed, 600°C/2h	7	Li-metasilicate, glass
Exposed, 600°C/2h + 725°C/2	9	Alpha-quartz, beta-spodumene, Li-metasilicate
No exposure, 800°C–850°C/2h	12–14	Alpha-quartz, Li-metasilicate, glass

Source: N.F. Borrelli and J.F. Schroeder, U.S. Patent Application, 2014.[14]

A number of these phases have low coefficients of thermal expansion (CTE); thus, by judicious choice of the thermal and exposure conditions it was found that one can produce a controlled mixture of a number of these phases with different values of CTE in order to produce a composite range from 3–14 ppm/C.

Table 3.10 shows a summary of the exposure/thermal treatments for the single composition that yields the listed values of the composite CTE, each containing a different mixture of the listed crystalline phases as indicated by the XRD, thus determining a composite value.[14]

The results shown in Figure 3.22a–c are a compilation of the CTE, the XRD spectrum, and the SEM measurement for three of the representative CTE values of 3 ppm/C, 7 ppm/C, and 14 ppm/C, respectively.

For the 3-ppm sample, the dominant phase is the low CTE is beta spodumene as indicated by the XRD. For the 14-ppm sample, the dominant phases are the high CTE LAS crystalline-stuffed alpha quartz and Li-metasilicate phases. Note that for the three intermediate CTE values, an exposure is required to produce the Li-metasilicate phase, which has a CTE of 7 ppm. By varying the amount of this phase by temperature, one can produce something higher or lower in CTE. It is possible in principle to produce a quasi-continuous CTE variation in a single piece using a combination of localized exposure and a thermal gradient in the heat treatment.

3.5.4 Other Variations

Until now, we have always the situation where Ag is the photonucleating agent; however, we noted in Chapter 2 that Au nanoparticles can be produced as well. It therefore makes sense to ask whether there is an Au-nucleated Fotoform. In Figure 3.23, this question is answered in a multitude of ways.

Table 3.11 lists the noble metal ion in a number of combinations with compositions. **K** has Ag and Au, **L** has less Au, **M** has Au only, **N** has less Au, **O** has Ag only, and **P** has low Ag, no Au. There has to be sufficient Au for the opal to form, and one can make FF with Au only and/or with low Ag without Au.

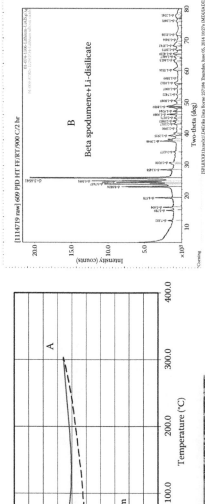

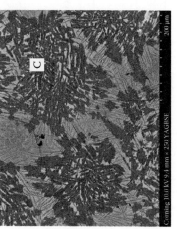

(a)

FIGURE 3.22 CTE (A), XRD (B), and SEM (C) data for the following values of the composite: (a) CTE = 3 ppm/C. *(Continued)*

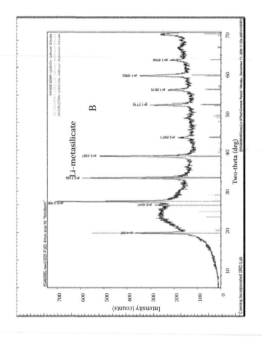

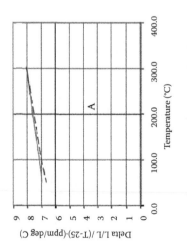

(b)

FIGURE 3.22 (CONTINUED)   CTE (A), XRD (B), and SEM (C) data for the following values of the composite: (b) CTE = 7 ppm/C.   *(Continued)*

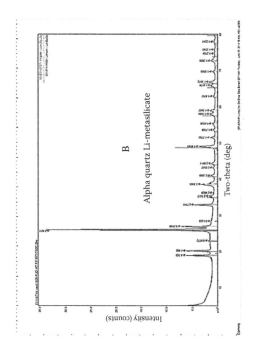

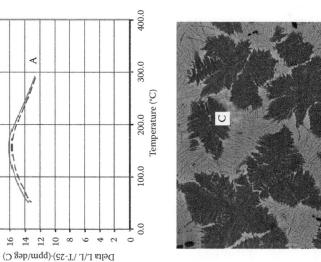

(c)

FIGURE 3.22 (CONTINUED) CTE (A), XRD (B), and SEM (C) data for the following values of the composite: (c) CTE = 14 ppm/C. (From N.F. Borrelli and J.F. Schroeder, U.S. Patent Application, 2014.[14])

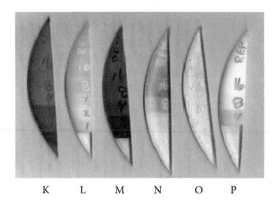

$$K \qquad L \qquad M \qquad N \qquad O \qquad P$$

FIGURE 3.23 Combinations of other nucleating noble metals whose compositions are provided Table 3.11.

TABLE 3.11
Glass Composition of the Glasses L-P Shown in Figure 3.23

(Mol%)	K	L	M	N	O	P
SiO_2	74.8	74.8	74.8	74.8	74.8	74.8
Al_2O_3	2.4	2.4	2.4	2.4	2.4	2.4
Li_2O	17.9	17.9	17.9	17.9	17.9	17.9
Na_2O	1.56	1.56	1.56	1.56	1.56	1.56
K_2O	2.54	2.54	2.54	2.54	2.54	2.54
ZnO	0.67	0.67	0.67	0.67	0.67	0.67
Sb_2O_3	0.08	0.08	0.08	0.08	0.08	0.04
CeO_2	0.006	0.006	0.006	0.006	0.006	0.006
SnO_2	0	0	0	0	0.007	0
Au	0.003	0	0.003	0	0	0
Ag	0.06	0.06	0	0	0.06	0
Au	0	0.0004	0	0.0004	0	0
Ag	0	0	0	0	0	0.0025

In Figure 3.24, we also show an SEM comparison of standard Ag/Au to some of the other nucleated versions.

The lack of color in the low Ag versions is due to the fact that there is not enough residual Ag^+ after forming the Ag nanoparticles; that is, all the Ag is used up in forming the nuclei. This residual Ag^+ is capable of being reduced during the thermal treatments. It is not generally known that exposed and developed Fotoform can produce a variety of tints as a consequence of this Ag reduction. This is shown in Figure 3.25. It is conjectured that this variety of tints must come from the Ag being reduced somehow on the Li-metasilicate phase where it is known that the different morphology of the Ag particle can produce different absorption features (see Section 9.2 in Chapter 9). Also notice that the Au-nucleated sample is rose-colored, which is consistent with the surface plasmon resonance absorption of Au (see the Appendix in Chapter 2).

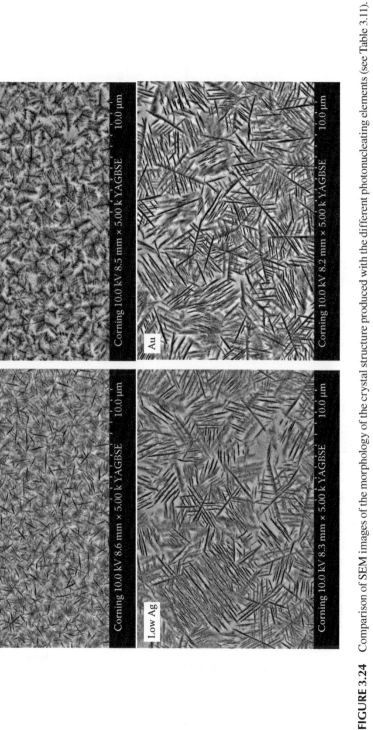

FIGURE 3.24 Comparison of SEM images of the morphology of the crystal structure produced with the different photonucleating elements (see Table 3.11).

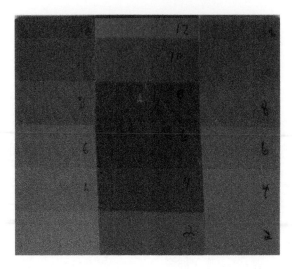

FIGURE 3.25 Various tints that can be produced in Fotoform as a consequence of the morphology of how Ag is reduced on Li-metasilicate crystallites.

Another version called Fotoceram™, is obtained if the exposed and thermally developed patterned material is heated to 800°C and the crystal phase converts to a Li-disilicate.

3.6 SECOND-STAGE NUCLEATION

3.6.1 NaF/Nepheline

The process discussed thus far in this chapter produces a crystalline phase from the glass nucleated by a photosensitively produced Ag nanoparticle. However, as was discussed in Section 3.2, there is another phenomenon that will be demonstrated here where the process continues through a separate second crystalline stage that was expressed in Equation 3.5 as a continuation of the process described by Equations 3.1 through 3.4. In this situation the initial photosensitive phase (Equation 3.4) is Fota-Lite; that is, NaF nanocrystals produced by the mechanism described above.[8] Now we describe a continuation of the process where a further nucleation occurs, producing an entirely new crystalline phase. The example is nepheline $(Na_3[AlSiO_4]_4)^4$ produced from the same glass that initially formed NaF upon heating to a higher temperature.[15] The glass composition in this case is a modified version of that of the Fota-Lite shown in Table 3.1 by substituting alumina for silica to increase the amount of the aluminate phase, and is shown in Table 3.12.

3.6.2 Exposure and Thermal Development

Typical exposure uses Hg or Hg/Xe lamps, producing UV lines in the range of 300–360 nm (within the range of the Ce^{+3} photoexcitation spectrum). The UV intensity is 5–10 mW/cm² with exposure times ranging from 5–60 minutes. The level of excitation is

TABLE 3.12

Composition of the Glass Where NaF Nucleates Nepheline (Na$_3$[AlSiO$_4$]$_4$)

Weight%	TSR
SiO$_2$	44
Al$_2$O$_3$	29.1
Na$_2$O	16.1
K$_2$O	6.1
ZnO	1.6
F–	3.3
Br–	1
CeO$_2$	0.02
Sb$_2$O$_3$	0.2
SnO	0.1
Ag	0.01

proportional to the energy; that is, intensity × time where higher intensity exposures will require less exposure times. Putting in these units, the range was ~3–36 J/cm^2.

The typical thermal treatments ranged from 600°C–700°C for hold times of 2–10 h. The desired optimum is to use a temperature where the unexposed region does not thermally produce the nepheline. The XRD results of an exposed and nonexposed sample both heated to 690°C are shown in Figure 3.26 and a treated sample of the exposed crystallized portion and the unexposed glass is shown in Figure 3.27.

In addition, samples were exposed with a tripled YAG laser (355 nm). Typical intensity for a 10-Hz rep rate was 2–3W, yielding a 0.6-J/cm^2 energy per pulse. The exposure times were 1–10 minutes using a 248-nm excimer laser with a pulse energy of 0.1 J/cm^2 for exposures intended only to produce the nepheline phase near the surface because of the depth of penetration of the 248-nm wavelength.

The advantages that may arise from the photo-patterned nepheline glass-ceramic (opaque to translucent) may be to utilize the different mechanical or optical properties from the patterned structure. In an appropriately patterned structure the body could have improved fracture toughness since controlled stresses can be developed between the exposed and unexposed regions due to the difference in thermal expansion.

The main significance of this study is a demonstration of a unique photonucleation phenomenon. The mechanism for this second stage of nucleation from the glass of a Na-aluminosilicate phase from a NaF nuclei that in turn was produced from a photonucleated Ag particle suggests no simple explanation in terms of what is generally dealt with in the literature for the standard nucleation phenomenon. What is clearly true is that the glasses in question are very unstable to spontaneous crystallization and any induced local fluctuation in composition could destabilize the total system. Experimentally it has been shown that Ag nanoparticles themselves do not nucleate the nepheline phase so the composition fluctuation they provide in developing the NaF phase is not sufficient to do anything further. In other words, it is the NaF phase that is the nucleus for the production of the nepheline phase.

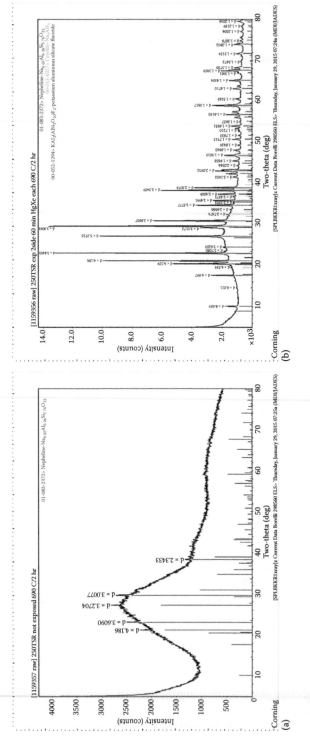

FIGURE 3.26 XRD trace of exposed and developed glass of the glass in Table 3.10: (a) unexposed and heated and (b) exposed and heated to 690°C. (From N.F. Borrelli et al., U.S. Patent Application, 2014.[15])

FIGURE 3.27 Exposed glass heated to 690°C whose XRD is shown in Figure 3.26.

3.6.3 LiF/Li-Aluminosilicate

The fact that one can produce a Na-aluminosilicate (nepheline) from an Ag photo-nucleated NaF suggests that if a photonucleated LiF phase could be produced in a similar way, then possibly one could produce a lithium-aluminosilicate in a manner similar to the Na-aluminosilicate in the NaF system. It was found that contrary to the photosensitive Fota-Lite glass described above where glass becomes opalized (forms NaF nanophase) in the exposed regions and remains clear in the unexposed regions, the LiF behaved in the opposite manner. The XRD measured from the unexposed area showed the presence of LiF and a slight amount of Li-aluminosilicate ($LiAlSi_2O_6$)[4] with nothing in the exposed region.

Based on this observation, the composition was altered to that shown in Table 3.13 by adding more alumina to enhance the LAS phase.[16]

For an exposure of 60 minutes at 350 nm at 10 mWcm[2] with a thermal development of 550°C/2 h, the XRD continued to show little crystal with a large glassy halo

TABLE 3.13
Glass Composition Leading to the LAS Phase after Heating to 550°C

SiO_2	75.68
K_2O	0.78
ZnO	6.03
Br–	0.71
Al_2O_3	7.22
CeO_2	0.04
Ag	0.03
Li_2O	7.23
Na_2O	0.03
F–	2.25

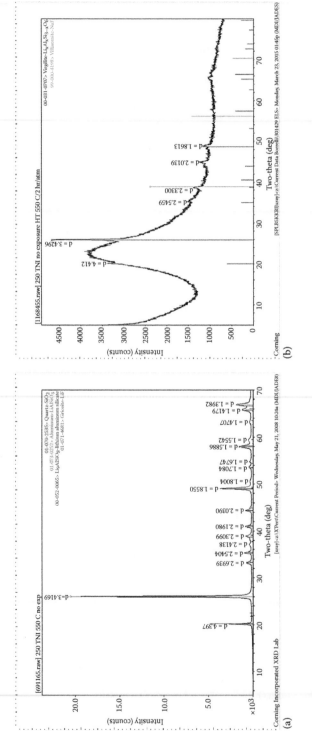

FIGURE 3.28 XRD of the glass exposed for 60 minutes at 10 mW/cm² and treated at 550°C/2 h: (a) nonexposed and (b) exposed.

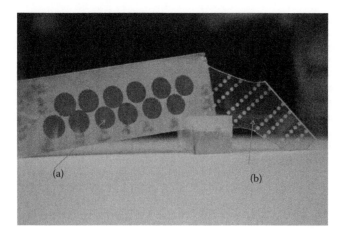

FIGURE 3.29 Image showing where the exposure light was blocked in the region that now consists of clear circles (a); and where the exposure light was covered by a mask with small open holes that are now opaque (b).

but now in the exposed region a well-crystallized condition was produced that was dominated by the stuffed beta-quartz LAS phase (see Figure 3.28).

The obvious conclusion is that the exposure in this case frustrates the formation of the crystalline LAS phase. The effect is more graphically shown in Figure 3.29a, which represents a situation where a mask made up of opaque dots is used to block the exposure light in the 1.5-mm circle areas, and in the reverse exposure mode (Figure 3.29b), the clear circular areas were exposed and the rest of the sample was not exposed.

The explanation of this effect is purely conjectural at this point. The conjecture is that the LiF indeed nucleates the LAS phase but does so in a spontaneous thermal manner that is not aided by light. In support of this explanation it should be pointed out that without the fluorine in the glass no LiF formed; one would have to heat the glass to a much higher temperature than 550°C to crystallize the LAS phase with the standard Zr nucleation.[3] In other words, the LiF can be an effective nucleating agent itself. For the exposed part the conjecture is that the formation of many tiny nuclei of LiF is not very effective in that there are too many nuclei that therefore frustrate growth. There is always some hint of the LAS phase in the exposed phase. It is nonetheless interesting to compare this behavior to the NaF/nepheline system discussed above where the exposure has the opposite effect.

REFERENCES

1. S.D. Stookey, *Journey to the Center of the Crystal Ball*, American Ceramic Society, Columbus, OH, 1985.
2. G.H. Beall, *Advances in Nucleation and Growth*, L.L. Hench and F.F Freiman (eds.), No. 5, American Ceramic Society, 1971.
3. G.H. Beall and D.A. Duke, *J. Mater. Sci.* 4, 340, 1969.
4. W. Holand and G.H. Beall, *Glass-Ceramic Technology*, Second Edition, John Wiley & Sons, Hoboken, NJ, 2012.

5. N.F. Borrelli, J.B. Chodak, D.A. Nolan, and T.P. Seward, *J. Opt. Soc. Am.* 69(11), 1514, 1979.
6. S.D. Stookey, *Ind. Eng. Chem.* 51, 805, 1959.
7. S.D. Stookey, *Ind. Eng. Chem.* 45, 115, 1953.
8. S.D. Stookey, U.S. Patent No. 2,684,911, 1954.
9. N.F. Borrelli, N. Lonnroth et al., U.S. Patent No. 9,011,720.
10. S.D. Stookey, G.H. Beall, and J.E. Pierson, *J. Appl. Phys.* 49, 5114, 1978.
11. N.F. Borrelli, D.L. Morse, R.H. Bellman, and W.L. Morgan, *J. Appl. Phys.* 24, 2520, 1985.
12. N.F. Borrelli, R.H. Bellman, J.A. Durbin, and W. Lama, *Appl. Opt.* 30(25), 3633, 1991.
13. N.F. Borrelli, *Microoptics Technology*, Second Edition, Marcel Dekker, New York, 2005.
14. N.F. Borrelli and J.F. Schroeder, U.S. Patent Application, 2014.
15. N.F. Borrelli et al., U.S. Patent Application, 2014.
16. N.F. Borrelli, G.H. Beall, and J.F. Schroeder, U.S. Patent Application, 2015.

4 Photorefractive Glasses

Now you see it and now you don't.

Magician's phrase

4.1 INTRODUCTION AND BACKGROUND

One of the more interesting phenomena that can be produced in certain glasses is the ability to permanently change the refractive index by exposure to light; if not directly, then with a subsequent thermal treatment. Patterns of the refractive index in glasses and crystals provide important optical devices such as diffraction grating, holograms, and gradient index lenses, to name the important ones; see Borelli[1] for a review. Unfortunately, there is an effect in certain ferroelectric crystals also called photorefractive, which has an entirely different origin stemming from an electronic effect that cannot exist in glass because it is an insulator. The refractive index of a material is defined by the polarization that is induced in the material in response to the electric field of the incident light. We can write the expression showing the connection to the refractive index in the followings way where the susceptibility χ is defined where we assume that χ and ϵ are real.

$$P = \chi E \quad \epsilon = n^2 = 1 + 4\pi\chi \tag{4.1}$$

From a microscopic point of view, one can discuss the polarizability of the ion or a bond and how it relates to the susceptibility simply by multiplying the susceptibility by the number of the ions or bonds;

$$\chi = N\alpha \tag{4.2}$$

In a simple-minded way, one can try to construct the refractive index of a glass as the sum of all the constituent concentrations of ions or bond/cm^3 multiplied by their respective polarizability. The anions other than the larger ones, such as Pb or Bi, have small values of the polarizability and therefore make little contribution to the refractive index so the refractive index is dominated by the oxygen polarizability in oxide glasses. Unfortunately, the polarizability is not really constant in all arrangements of the glass structure. For example, bridging and nonbridging oxygens have a different polarizability than oxygens bonded to boron or aluminum rather than silicon. Nonbridging oxygens are mentioned in a number of the chapters throughout the book relevant to the role they play in photosensitivity and photochromism. As we will see, there are two apparently distinct origins of induced refractive index change or photorefractive effects in glass. In one example, the structure of the glass itself has been modified by the light by the breaking of bonds and the subsequent atomic rearrangement gives rise to the refractive index change; that is, the polarizability of the constituent ions or atoms has been altered in the new structure. In this situation there will always be an accompanying induced absorption as a result of the structure change. The proposed mechanisms for the various examples of induced refractive index change may have more detailed models of optically induced altered atomic structure changes. The Kramers-Kronig expression relates the real part of the refractive index n to the induced absorption or imaginary part of the refractive index, k. The expressions one usually sees are shown in Equation 4.3 in terms of the complex dielectric constant that defines the complex refractive index.

$$
\begin{align}
&\text{(a)} \quad D = \epsilon E \\
&\text{(b)} \quad \epsilon = \epsilon_1 + i\epsilon_2 \\
&\text{(c)} \quad \epsilon_1 = 2nk \quad \epsilon_2 = k^2 - n^2 \\
&\text{(d)} \quad n = n + ik
\end{align}
\tag{4.3}
$$

The main point of the Kramers-Kronig expression is that it relates the real part of the refractive index to the imaginary part that is the absorption through the following dispersion integral:

$$
n(\omega) - 1 = \left(\frac{1}{2\pi}\right) \int_0^\infty \frac{\omega' k(\omega') d\omega'}{\omega^2 - \omega'^2}
\tag{4.4}
$$

The point here is that any optical change induced by the exposure without regard to any specific origin will result in a related refractive index change, as shown in Equation 4.4, so in reality the Kramers-Kronig expression is not a description of a physical mechanism of photorefraction phenomenon but a necessary result.

However, there is another photorefractive mechanism more often called photothermal because it requires a thermal treatment after the light exposure to produce the refractive index change. Here invariably the induced index arises from the production of another phase within the glass. This was discussed in Chapter 3 in some detail for certain glasses, such as Fota-Lite, for example. In this chapter we are concerned with the refractive index change induced by the light in these materials. Although it is not obvious this also follows Equation 4.4 as we see that the absorption edge in these induced phase separated glasses does shift correspondingly.

We will first look at two of the photothermal examples, the first is based on the development of the NaF phase in the exposed region as the source of the index change. Other properties of this Fota-Lite material were discussed in Chapter 3. The second example is more complex in that although it is still based on a thermally developed phase—Ag halide in this instance—the index change is produced after the phase is formed directly by exposure to light; that is, thermal treatment is necessary. The third case arises from a different mechanism much more closely described by Equation 4.4. It was first observed in Ge-doped silica optical fibers[2,3] and the effect was termed fiber Bragg grating (FBG) as a consequence. The example described here will be of that produced in the bulk glass version of the multicomponent alkali germane-silicate glass. As we will see, this comes in two different versions, one is produced in as-made glass and the other is aided and enhanced by the addition of molecular hydrogen.[4,5] The photorefractive effect in H_2 loaded aluminosilicate glasses will also be shown. The method in the fourth case is not quite photosensitive in the same sense as the others although light is used to produce the index change. This method produces unique effects as a consequence of a laser that produces femtosecond pulses. We will only briefly cover this method for completeness since it does not seem to depend too strongly on glass composition.

4.2 PHOTOREFRACTIVE EFFECTS IN NaF-BASED GLASSES

In Chapter 3, we introduced and discussed the photosensitively produced NaF nanophase glass system; there we dealt with the optical processes leading to the formation of the NaF nanophase together with examples of the patterns and colors produced by the photo-produced phase. The NaF particle size was large enough or the particle concentration high enough to produce significant Raleigh scattering to yield the opal appearance when exposed and thermally developed. However, with slight modification of the composition (three times more Ag) but more importantly utilizing a much higher exposure level and a significantly different thermal development schedule, one can reduce the size of the NaF particles to a point where the resulting glass is relatively transparent after exposure and thermal treatment. The explanation for why these differences result in a more transparent condition is that the higher energy exposure coupled with a thermal schedule produced smaller NaF nano-sized particles. The refractive index of the NaF phase is much lower (1.32) than the surrounding glass (1.5) and this combined with high transmission (low scattering because of the small NaF grain size) allows for a high spatial resolution medium suitable for producing a wide variety of photorefractive effects. Table 4.1 lists the glass composition.

TABLE 4.1
Composition of Photosensitive
NaF-Based Glass

Weight%	THA
SiO_2	66.9
K_2O	0.75
ZnO	6.5
Na_2O	16.3
Ag	0.034
Br^-	1.26
F^-	2.5
Al_2O_3	6.5
CeO_2	0.037

4.2.1   Exposure and Thermal Treatment

The exposures are normally done with either a tripled Nd/YAG laser operating at 355 nm or a continuous wave 320 nm He-Cd laser, although Hg UV lamps have been used as well. The exposure is essentially energy-determined so the exposure is proportional to the amount of the light absorbed (αI) × the exposure time, hence (αTt). Each source has a controlled intensity and the absorption at that wavelength is determined by the material. The point is that in general, other than the physical depth of the exposure and the length of the exposure, the wavelength of the exposure source is not important. We will see there are some exceptions to this rule. The wavelength of all these sources are well suited to the wavelength of the absorption of the Ce^{+3} photosensitizer, as shown in Figure 4.1.

For many of the applications, a coherent light source provided by a laser is required to be able to create interference patterns and thus allow holographic images to be produced through the induced refractive index change. In most cases presented here, a fused silica phase mask[1] was used to provide the interference pattern. The 355 nm laser light beam was made incident on the phase mask that was in intimate contact with the photosensitive glass. The thermal treatment schedule to develop the NaF phase was done at a low enough temperature to keep the glass clear; that is, to minimize light scattering. The difference in composition between Fota-Lite as given in Chapter 3 and the composition listed above for the photorefractive glass together with the thermal development schedule was primarily aimed at being able to keep the NaF nanophase particle size small and thus maintain a high degree of transparency. The loss is dominated by scattering. Figure 4.2 lists the typical exposure conditions and provides a diagram of the thermal treatment schedule.

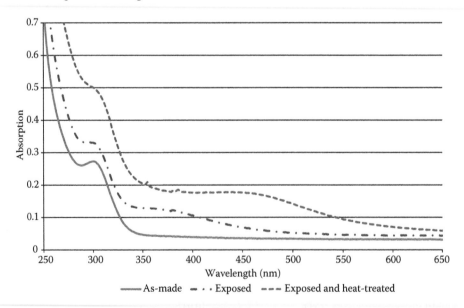

FIGURE 4.1   Absorption spectra of the photosensitive glass after various stages of the treatment (initial, exposed, and heated), as noted on the graph.

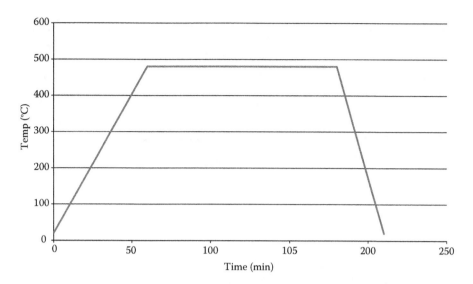

FIGURE 4.2 Thermal development schedule for NaF-based photorefractive glass. Typical exposure conditions are pulsed 355 nm laser 2.5 Wcm², 10 Hz for 5–20 minutes (~150 J/cn²).

This schedule should be compared with the one shown in Figure 3.3 in Chapter 3, for Fota-Lite to see the significant difference between the two. If the glass composition of Table 4.1 used this exposure and was subjected to the thermal schedule of Figure 3.3, it would be opaque. On the other hand, if the Fota-Lite composition in Table 3.2 were treated as noted in Figure 4.2, it would be clear and show about one-third the induced refractive index change compared to the composition shown in Table 4.1. This latter point would indicate that in addition to the higher Ag content in the photorefractive version of the Fota-Lite composition, the exposure conditions themselves have an effect on how much and what nanoparticle size of the NaF phase will develop. Putting it more succinctly, the exposure condition/thermal development does influence the ultimate NaF nanoparticle size and the number density that influences the magnitude of the ultimate Δn. Clearly, the glass composition is not different enough to account for the different photorefractive outcomes. This offers a great example of the complexity of the way light can interact with glass. Glasses need not be a static material, in this case manifesting anywhere from an opaque white, a variety of colors, or clear photorefractive material with essentially the same glass.

4.2.2 Characterization of the Photorefractive Effect

The quantitative characterization of the induced refractive index effect is done by an interferometric measurement. It measures the phase shift in 633 nm waves produced by the exposure area relative to the unexposed area. The induced refractive index is then determined by the following expression where ϕ is the phase, λ is the wavelength of light (0.633 nm), and t is the sample thickness:

$$\Delta n = \Delta \phi \lambda / t \qquad (4.5)$$

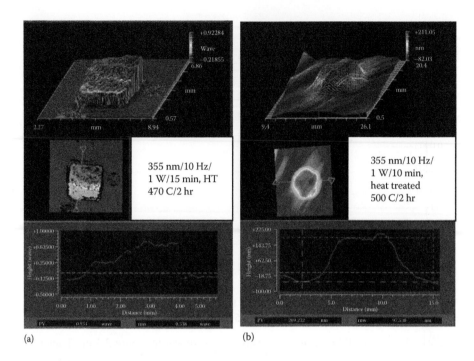

(a)   (b)

FIGURE 4.3 (a) ZYGO interferometer measurement output of induced refractive index change in glass composition from Table 4.1; exposure was with a pulsed 355 nm laser 1 W/cm^2 at 10 Hz for 15 minutes. From the measured 633 nm phase shift using Equation 4.5, the induced $\Delta n = 0.6 \times 10^{-3}$; (b) ZYGO measurement of the Fota-Lite composition of Table 3.2 (previous chapter) exposed under the same conditions as those listed in (a), $\Delta n = 0.26 \times 10^{-4}$ (see discussion above).

Figure 4.3 shows a typical interferometric (ZYGO) measurement result of the phase shift after exposure and thermal treatment of a 1-cm^2 pattern. Using results from this type of measurement one can obtain the induced refractive index as a function of 355 nm exposure time, as shown in Figures 4.3 and 4.4.

Maintaining optical loss as low as possible is an important feature of the performance in any proposed application. A typical loss example, loss versus wavelength data, is shown in Figure 4.5.

A hologram fabricated in this glass utilizing an exposure through a phase mask is shown in Figure 4.6. What this result demonstrates is the spatial resolution that this glass is capable of recording in the neighborhood of 0.5 μm. One of the major commercial applications of this holographic capability is for making frequency stabilizing filters for laser diodes.

4.3   PHOTOREFRACTIVE EFFECT
IN Ag HALIDE-CONTAINING GLASSES

A number of years ago, an Ag-containing glass was invented at Corning by Wm. Armistead and worked on some time later by D.L. Morse (unpublished), which

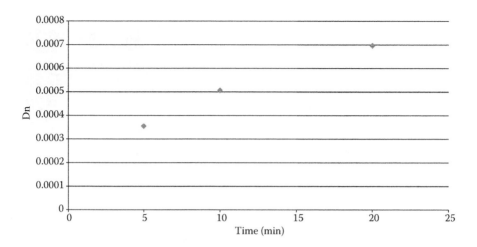

FIGURE 4.4 Induced refractive index as a function of exposure time, 1 W/cm², 355 nm 10 Hz.

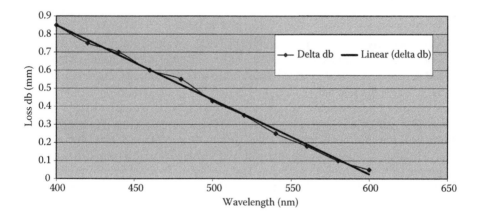

FIGURE 4.5 Measurement of the loss as a function of the wavelength of the photorefractive material after UV exposure and thermal treatment.

was called rainbow glass.[6] The name arose from the fact that different colors could be produced depending on the time and temperature of a postthermal treatment time at 600°C. A patent was issued dealing with some aspects of this effect[6]; however, the results were never published in the open literature. Figure 4.7 shows an example of such a glass subjected to a thermal gradient.

In addition to the presence of Ag, the glass composition also has to contain Cl and Sb_2O_3 for the effect to be observed. Although this coloration is not produced by exposure, nonetheless the similarity of the color formation, albeit thermally to that of the two NaF-based polychromatic glass and Fota-Lite glass discussed in Chapter 3, suggested some common mechanism involving Ag halide as the host for the coloration in this case rather than NaF. Although the formation of Ag halide

FIGURE 4.6   Holographic filter made in the photorefractive glass using an exposure through a phase mask; 1 W/cm^2, 355 nm, 10 Hz.

was never actually verified experimentally by XRD, the prevailing opinion was that because the glass contained Cl (see for example Table 4.2), the Ag halide phase was formed during the thermal treatment (see Figure 4.8). The amount and/or size of the nanocrystals were too small to detect by XRD. It was thought that this Ag halide phase ultimately produced the reduced silver on/or in the Ag halide particle. In other words, the origin of the colors was attributed to the absorption of the Ag nanoparticle arising from the state of completion of the reduction process of the halide particle to silver. The longest times of thermal treatment did indeed lead to a color (yellow) that corresponded to the absorption of spherical nanoparticle silver (see Section 2.5.2). XRD analysis did yield the silver diffraction peaks in the exposed region, as can be seen in Figure 4.8.

It was hoped that the Ag halide phase would show up in the unexposed region, but apparently, the crystal size is too small. These rainbow glasses were not thought to be photosensitive in the sense discussed in Chapters 2 and 3. It was found that exposure to 248 nm light had sufficient energy to create electron/hole pairs leading to the reduction of Ag in or on the Ag halide particle.[7]

This produced the absorption involving Ag through a subsequent thermal treatment at a temperature lower than that where the silver would typically be thermally reduced. So, the conjecture was that the Ag nanoparticles formed on the thermally developed Ag halide phase. This conjecture was confirmed, as shown in Figure 4.9 in the TEM images of the AgCl phase as well as Ag.

Moreover, the absorption spectrum of the exposed region clearly shows the presence of nanoparticles of Ag, which will be shown in the next section.

A problem was to produce a sufficient transmittance at the 248 nm excitation wavelength to produce electrons to any reasonable depth into the glass. The UV transmittance curves are shown in Figure 4.10.

FIGURE 4.7 Photograph of rainbow glass made by applying a heat treatment utilizing a thermal gradient to the glass with the composition given in Table 4.2.

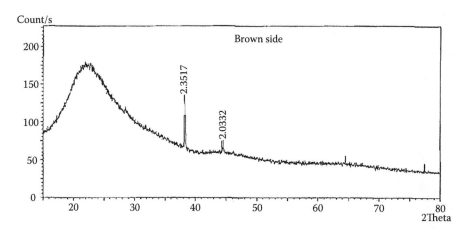

FIGURE 4.8 XRD of exposed photosensitive rainbow glass of the composition listed in Table 4.2, showing the two lines characteristic of AgCl.

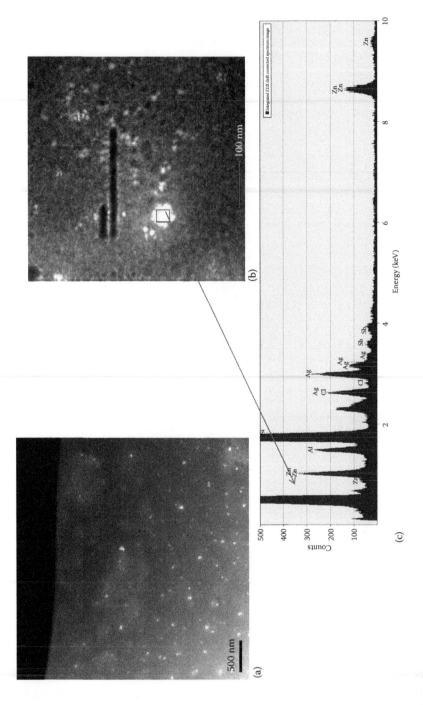

FIGURE 4.9 TEM images of exposed and developed rainbow glass showing the presence of both AgCl and Ag: (a) images of the AgCl, and (b) the spot that was elemental analyzed with results shown in (c).

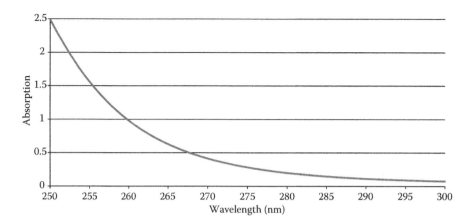

FIGURE 4.10 Absorption spectrum of a 1-mm thick sample of rainbow glass to indicate that the excitation at 248 nm is strongly absorbed, thus limiting the depth of the refractive index modification to ~0.2 mm (see Figure 4.11).

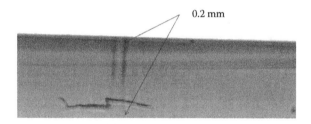

FIGURE 4.11 Cross-section micrograph of exposed rainbow glass showing the depth of the effect limited 0.2 mm because of the required excitation wavelength at 248 nm.

The main composition components that act as absorbers controlling the UV edge are the Sb and the Ag, both of which are required for the photorefractive effect. The composition was optimized to make the glass as transparent as possible while still maintaining a large induced refractive index. This composition is shown in Table 4.2. One can never achieve a value of the transmission at 248 nm, which is much higher than that allowing a photosensitive response to a depth of only 0.20 mm of the sample, as shown in Figures 4.10 and 4.11.

4.3.1 EXPERIMENTAL RESULTS

The following results for photorefractive Ag halide-based glass are provided below. The optical absorption spectrum of the as-made glass compared to the thermally treated exposed and unexposed glass are shown in Figure 4.12. The strong absorption feature in the 450–500 nm spectral region clearly corresponds to that of the surface plasmon resonance of Ag, as discussed in Section 2.5. The 248 nm excimer laser-induced refractive index results are shown in Figure 4.13 along with its dependence on exposure time.

TABLE 4.2
Composition in Weight% of
Photorefractive Rainbow Glass

(Weight%)	I
SiO_2	67.1
B_2O_3	16.7
Ag	0.33
Cl^-	0.22
Na_2O	4.3
Na_2O	3
F^-	1.7
Sb_2O_3	1
ZnO	5
Al_2O_3	3

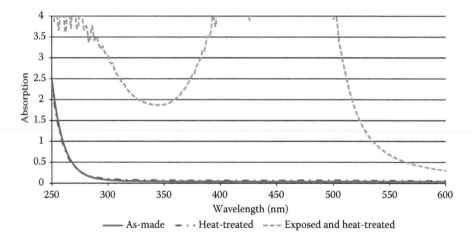

FIGURE 4.12 Absorption spectrum of the exposed and thermally developed Ag halide glass compared to the unexposed glass. The strong absorption feature ~500 nm is attributed to the surface plasmon absorption of Ag nanoparticles.

The glass was found to be quite sensitive showing essentially saturated behavior after 60 s exposure at 248 nm. It was found that a heat treatment of 560°C for 1.5 h was sufficient to produce the Ag phase in the exposed regions and no color in the unexposed region. The data in Figure 4.14 shows the writing of a phase grating using a mask. The mask exposure produces a quasi-sinusoidal varying refractive index change that constitutes a diffractive phase grating. The pattern shown in Figure 4.14a is observable by a slight absorption difference between the exposed and unexposed regions whereas that in Figure 4.14b shows the index pattern contrast by viewing by a dark field technique. In Figure 4.14c, we can see the diffraction pattern produced by the phase grating by making a HeNe laser incident on the grating. The brightness

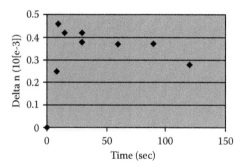

FIGURE 4.13 Induced refractive index as a function of exposure time with 248 nm, 15 mW/cm² 10 Hz. (From N.F. Borrelli, G.B. Hares, and J.F. Schroder, U.S. Patent Pending. With permission.[7])

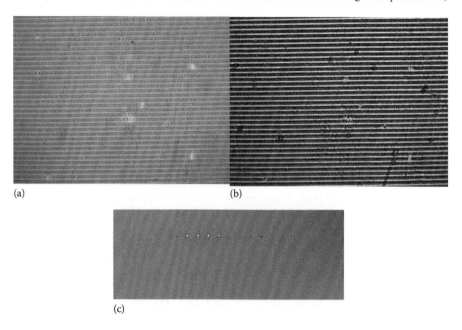

FIGURE 4.14 Phase grating produced in rainbow glass using a 248 nm exposure through a mask: (a) shows the grating seen by a slight absorption difference between the exposed and nonexposed regions; (b) shows the image of the refractive index difference; and (c) shows the diffraction pattern produced by the grating.

of the spots is determined by the diffraction efficiency of such a grating that follows a Bessel function form,[1] where j is the order of the spots.

$$\text{Efficiency} = J_j^2(2\pi\Delta nT/\lambda) \tag{4.6}$$

T is the thickness of the grating, Δn is the index contrast, T is the thickness, and λ is the wavelength of light.

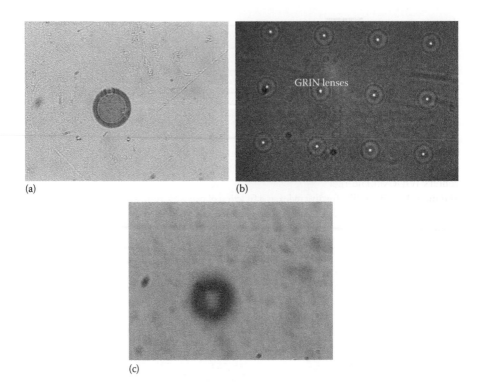

FIGURE 4.15   GRIN lens process: (a) the exposed area obtained by exposure through a pin-hole, (b) image of the focal plane of the lens, and (c) image of a pattern from a single lenslet. (From N.F. Borrelli, *Microoptics Technology*, Marcel Dekker, New York, 2005.[1])

Another manifestation of the induced refractive index effect was to make a gradient index lens by exposing the glass through a small ~1-μm opening. Because of the diffraction of the light as a consequence of the small opening, a quasi-Gaussian intensity exposure pattern is produced, which is a decent approximation of a parabolic intensity exposure. To the extent that the induced refractive index is linear with the intensity (energy = intensity × time), then the refractive index will have a quasi-parabolic radial dependence that is required to make an imaging gradient-index (GRIN) lens.[1] The reader is referred to Borrelli and Borrelli et al.[1,7] for a detailed optical description of the GRIN lens. Figure 4.15a through c shows an example of a photorefractive fabricated GRIN lens. The estimated focal length of the lens is 0.3 mm, corresponding to an NA of the order of 0.2 and a $\Delta n \sim 10^{-3}$.

4.4   LIGHT-INDUCED REFRACTIVE INDEX CHANGES IN GLASSES

In the above description of the photorefractive phenomenon we dealt with special glasses, meaning glasses where the photorefractive effect is derived from the development of a secondary phase. However, there is also the possibility of using light with sufficient energy ($h\nu$) to induce structural changes in the glass network itself that would lead to refractive index changes. These are cases where the intensity of the exposing light makes a difference contrary to the above discussion of the photo-refractive effect being dependent only on the energy. An example of this is shown in Chapter 7, which deals with induced absorption from UV exposure in fused silica. In this particular example the index change is generally attributed to an induced density change but there is also a hint that some of it is indeed photorefractive[8] in origin. This means that the light has altered the structure in such a way that the polarizability of certain bonds, or arrangement thereof, has been changed.

The initial topic will be a variation on what is known as fiber Bragg grating (FBG). This was an effect first observed in Ge-doped silica optical fibers[2,3] where UV exposure led to optical loss associated with certain defects called oxygen-deficient centers (ODCs). The reaction of the ODCs with molecular hydrogen was found to enhance the effect. This effect and its origin will be discussed in more detail in the next section. The strong induced absorption in the UV region produced a refractive index change out into the visible and NIR region through dispersion. This effect then was used to pattern expose the fiber to produce diffraction gratings using 248 nm excimer radiation.[1] The aspect to be covered here originates from the discovery that this UV-induced effect involving Ge could be also produced in bulk glass. This will be the topic of the next section. It will not only be the effect produced in Ge-doped glasses but also the effect in more conventional aluminosilicate glasses where deep UV exposure in hydrogen-loaded glasses show measurable photorefractive effects. To conclude, there will a discussion of the effect of 800 nm fs-laser exposure in producing significant refractive index changes. All of these effects derive from a mechanism where the glass structure is altered in some way, leading to a refractive index change.

4.5   BACKGROUND ON Ge-CONTAINING GLASSES

We will briefly review the phenomenon as it was seen in GeO_2-SiO_2 optical fiber preform canes. The preform cane is the core/clad structure before the draw into optical fiber. There are two manifestations of this photorefractive effect in this material having two distinct mechanisms, one involving an oxygen-deficient center[9,10] and the other involving a reaction with light when molecular H_2 is present.[4,5] The relevant effect pertaining to the discussion below concerning the effect in photorefractive Ge bulk glass has to do with the molecular H_2-initiated effect so we only will briefly review this aspect as an introduction.

It was found that a much larger photo-induced refractive index change could be produced in GeO_2-SiO_2 compositions by impregnating the fiber with molecular hydrogen before the exposure.[4,5] This impregnation is carried out at or slightly above room temperature in order to ensure that no reaction occurs. The pressure of the hydrogen is high to allow the order of 10^{20} molecules/cm^3 to be dissolved. The mechanism is thought to involve the direct attack of the Si-O-Ge bond with molecular hydrogen in the presence of the radiation[8] leading to both GeH and SiOH species, both of which are observed in the NIR spectra. The increase in the absorption in the deep ultraviolet as a consequence of the exposure is large, as shown in Figure 4.16. Note that the magnitude of the photo-induced absorption in the hydrogen loaded preform slices is independent of the initial defect absorption at 240 nm. The latter is the result of the consolidation of the preform being done in a O_2 ambient rather than the

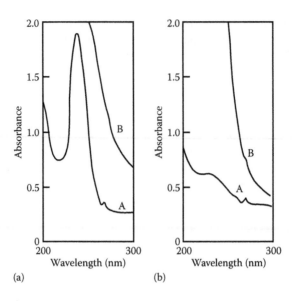

FIGURE 4.16 UV and VUV absorption spectra of H_2-loaded 7% GeO_2 preforms (a) before (curve A) and after (curve B) 248 nm exposure in the case where the ODC is clearly present; and (b) the case where the ODC is not present. The measured induced refractive index is essentially the same. (From N.F. Borrelli, *Microoptics Technology*, Marcel Dekker, New York, 2005.[1])

normal He ambient. The important point is that the amount of induced absorption is independent of these different conditions. This speaks to the mechanism of the H_2-loaded case as being different than the induced index effect in the H_2-free situation where the ODC was required. This is relevant to the discussion in the following section where we reproduce the photorefractive effect in a conventionally melted bulk glass composition where the ODC does not form.

4.5.1 H_2-Loaded Silicogermanate Glass

The intent of this research effort was to produce a glass that could be made by conventional melting and forming techniques. In addition, such a glass would possess comparable excimer laser-induced photosensitivity to that of the above mentioned hydrogen-loaded germania-based optical fibers that formed the basis for what had become the standard for fabricating fiber Bragg grating. In other words, reproduce the photorefractive behavior in a bulk glass. The general compositional approach was to use a glass with good intrinsic UV transmission, in particular in the vicinity of the intended exposure wavelength of 248 nm. The glass that was selected as the basis for the experiment was the glass used for germicidal lamp envelopes because of its good 256 nm transmission. The next step was a partial substitution of the required germania component for silica to the concentration of maintaining good transmittance at 248 nm.[11] The general effect of the substitution of germanium for silica is a wavelength red (longer wavelength) shift of the fundamental edge (see Figure 4.17).

This wavelength cutoff against the desirable effect of higher germania content had to be balanced in terms of the magnitude of the induced refractive index change since it is the Ge-O-Ge or Si-O-Ge bond that is responsible for the photorefractive effect similar to the observed effect in the fiber. The composition is given in Table 4.3, and the measured physical and thermal properties are provided in Table 4.4.

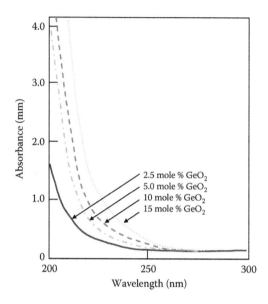

FIGURE 4.17 Absorption spectra of Ge-based photorefractive glass composition as a function of GeO_2 content. (From A. Streltsov and N.F. Borrelli, *Opt. Lett.* 26(1), 42, 2001.[14])

TABLE 4.3

Composition of Ge-Based Photorefractive Glass

Mol%	1
SiO_2	62.5
GeO_2	2.5
Na_2O	2.5
Al_2O_3	2.5
B_2O_3	30
H_2 ($\times 10^{19}$)	5
Δn (mod, $\times 10^{-4}$)	2.8

TABLE 4.4

Physical and Thermal Properties of Glass Composition Given in Table 4.3

		Viscosity Temperatures		Liquidus Temperatures	
Softening point	664°C	100 poise	1668°	Air	1090°C
Annealing point	430°C	1000 poise	1342	Internal	1090°C
Strain point	391°C	10,000 poise	1112	Platinum	1080°C
Expansion	46	100,000 poise	941	Phase cristobalite	
Density	2.353 gm/cc	1,000,000 poise	809		

4.5.2 INDUCED REFRACTIVE INDEX

The samples were loaded with molecular hydrogen at a pressure of 100 atm in a pressure vessel at 200°C. For the thickness of 1 mm, the diffusion time to achieve 95% of the ambient at the midplane was approximately 3 days. The temperature of the loading was limited to well below the temperature at which the H_2 reacts with the glass, as evidenced by an increase in the SiOH and GeH species. The molecular hydrogen concentration was determined from the magnitude of the calibrated IR absorption peak at 4126 cm^{-1}, which corresponds to the hydrogen stretching vibration.

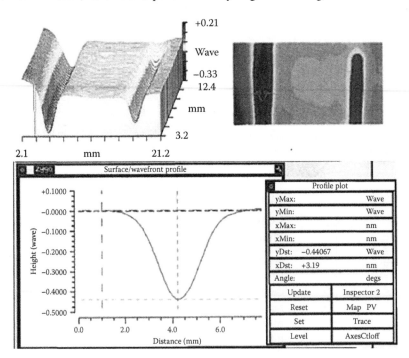

FIGURE 4.18 ZYGO interferometer measurement results. The phase shift is shown in a 3-D format in the upper-left image, in a graphical form in the lower-left image, and is numerically tabulated in the box under yDst = 0.44 waves (633 nm). (From N.F. Borrelli, C.M. Smith, and V. Bhagavatula, *Proc. SPIE* 4103, 242, 2000. With permission.[11])

The method used to measure the induced refractive index was interferometry (ZYGO). A typical example of the induced refractive index derived from this measured phase distortion in 633 nm waves, Δ (nL) is shown in Figure 4.18. It shows the induced phase shift in a typical interferogram display from which the induced phase shift data was obtained. Figure 4.19 shows the induced absorption produced by the exposure that leads to the refractive index change. Figure 4.20 shows the dependence of the induced refractive index change and the square of the exposure fluence.

A definitive test of the utilization of this phenomenon in a device application was to write a diffraction grating into the photorefractive glass in the manner shown in the inset in Figure 4.21. A spatially filtered 248 nm beam from a KrF excimer laser was incident on a phase mask in contact with the large face of the glass as shown. Spatial

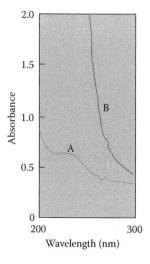

FIGURE 4.19 Absorption curves measured before (curve A) and after (curve B) 248 nm exposure in an H_2-loaded sample of the glass presented in Table 4.3 showing the large induced change that led to the index change; $H_2 = 10^{20}/cm^3$. (From N.F. Borrelli, *Microoptics Technology*, Marcel Dekker, New York, 2005.[1])

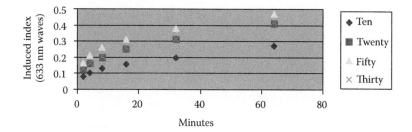

FIGURE 4.20 Induced 633 nm wavefront change for the silicogermanosilicate glass as a function of exposure time with 248 nm exposure time (90 mJ/cm² 50 Hz) for four levels of H_2 loading (atm). (From N.F. Borrelli, C.M. Smith, and V. Bhagavatula, *Proc. SPIE* 4103, 242, 2000. With permission.[11])

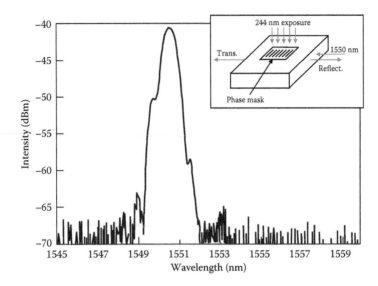

FIGURE 4.21 Inset: Schematic of the exposure format with the phase mask placed in contact with the face through which a 248 nm exposure was made. The graph shows the measured grating performance centered at 1550 nm.

filtering was done to reduce the number of modes of the laser to increase the spatial coherence so that the phase of the two interfering beams from the phase mask would be maintained through the 1-mm thick sample. The graph in Figure 4.21 is a plot of the 1550 nm light measured in decibels passing through the sample perpendicular to the grating, as shown in the inset. This is a Bragg grating 1550 nm filter with a 40-db rejection ratio. The depth of the grating in cross section is shown in Figure 4.22.

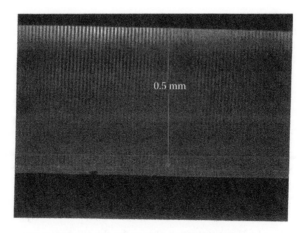

FIGURE 4.22 Cross section of grating described in Figure 4.21.

4.5.3 MECHANISM

There is quite a bit of literature on the nature and origin of the photorefractive effect dealing with fiber Bragg grating both for H_2-loaded and nonloaded mechanisms, as reviewed in Borrelli.[1] As mentioned, the nonloaded mechanism involves an oxygen vacancy center that produces strong absorption at 240 nm. The 248 nm exposure is claimed to produce two Ge centers and the new absorption is a result of these centers. Since the mechanism for the H_2-loaded effect is different, as we will see, the reader is referred to in References [8–11] for a detailed description of this case.

The H_2-loaded mechanism is shown in Figure 4.23 where the action of the 248 nm light is to break the Si-O-Ge bond whose fragments react with the molecular hydrogen to produce GeH and SiOH color centers, the absorption of which is indicated in Figure 4.24. We seem to have reproduced this mechanism in conventionally melted glass with the composition given in Table 4.3 with an attendant induced refractive index change.

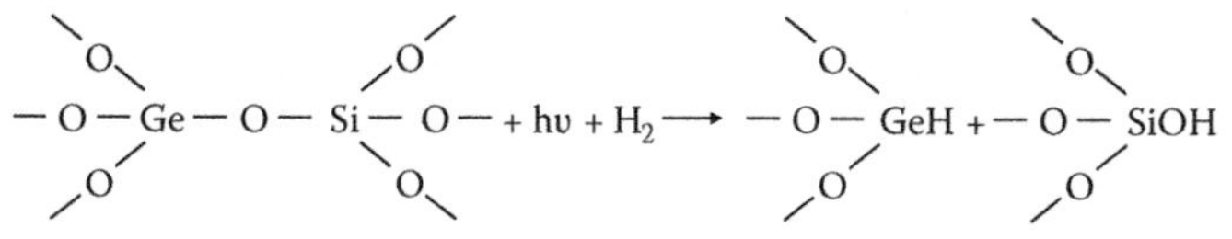

FIGURE 4.23 Schematic representation of the bond scission produced by 248 nm light and the subsequent reaction with molecular hydrogen.

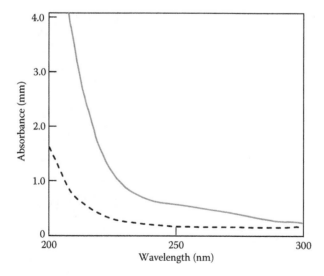

FIGURE 4.24 Induced absorption in germanosilicate glass produced by 248 nm exposure leading to a refractive index change.

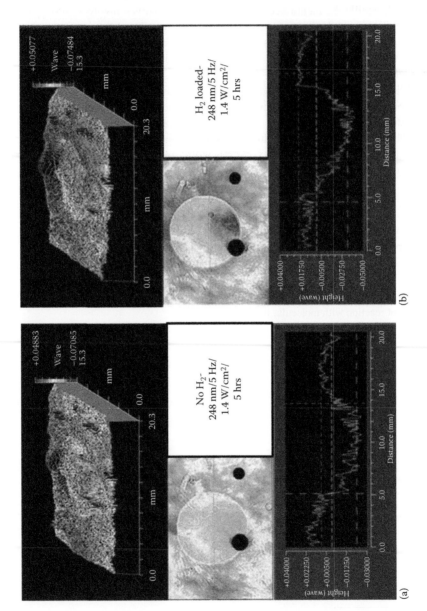

FIGURE 4.25 Interferometric result for 248 nm exposure (1.4 W/cm^2 5 h, 10 Hz) of an alkaline-earth aluminosilicate glass: (a) no loading with $\Delta n = 0.33 \times 10^{-4}$ and (b) with H$_2$-loading $\Delta n = 0.5 \times 10^{-4}$.

4.6   OTHER GLASS COMPOSITIONS

4.6.1   BACKGROUND

The previous section dealing with H_2-loaded germanosilicate glasses produced significant 248 nm induced refractive index changes. This brought about the possibility of looking for the photorefractive behavior of H_2-loaded effects in more common commercial glasses such as alkali and alkaline-earth aluminosilicates. In other words, would the same mechanism of H_2 reacting with deep UV light, as schematically expressed in Figure 4.23, be operative in silicate glasses with and without Ge? We will see in Chapter 7, all oxide glasses develop color centers of one kind or another under deep UV exposure, so the question presents itself as to whether they would react with H_2 in a similar manner and produce refractive index changes.

4.6.2   EXPOSURE RESULTS

Exposures were done without a phase mask and the induced index change was measured interferometrically (ZYGO). The 240 nm excimer exposures were all for 4 h. All phase mask exposures were done with a spatially filtered 248 nm laser to increase spatial coherence. In the latter case, Δn was computed from induced diffraction grating. Exposure time was 25 minutes.

Figure 4.25 shows the ZYGO results for the exposure of an alkaline-earth aluminosilicate glass not H_2-loaded (a) and H_2-loaded (b). There is an induced index change even in the not-H_2-loaded case, apparently from the induced absorption, as shown in Figure 4.26. An approximately 2× increase was obtained in the Δn between H_2 and the as-made; however, a much lower increase was achieved in the SiO_2GeO_2 system.

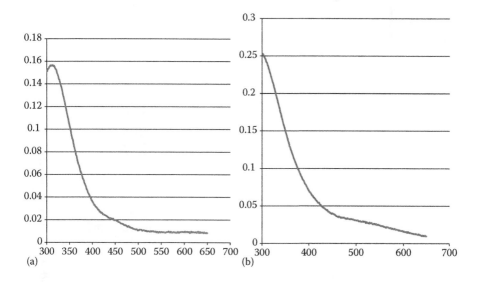

FIGURE 4.26   Induced absorption for the two samples shown in Figure 4.25. (a) Delta Abs versus wavelength; H_2-loaded; (b) Delta Abs versus wavelength; and as-made.

4.7 FEMTOSECOND LASER-INDUCED INDEX CHANGES

4.7.1 BACKGROUND

Since the introduction of mode-locked 800 nm Ti-sapphire lasers over the last decade that were able to produce laser pulses in the <100-fs range, there has been much work published dealing with induced refractive index changes and optical structures derived from such changes.[12–14] Because of this prior available information the discussion here will be limited since other sources are available that cover the topic in much more depth. It is included here not only to complete the discussion of the photorefractive phenomenon in glass but also to point out the distinct difference in the fs-laser index change mechanism compared to those discussed above. The differences are the following:

1. The index change is a function of *intensity* not energy; intensity can be as high as 10^{12} W/cm^2
2. It is a multiphoton effect so a pattern can be written inside a body by using a suitable focusing lens
3. The pulse duration is shorter than the thermalization time
4. There is little dependence of the effect on the glass composition ranging from silica to soda-lime to sulfides

As a consequence the photorefractive mechanism is likely quite different from an energy-derived mechanism as covered above. In those cases, the refractive index change arises from how many species or structure alterations are produced that correspond to the energy per unit volume delivered. The next discussion will show the fs-laser exposure methods and some examples of the laser written waveguide pattern in a number of different patterns followed by a brief discussion of the suggested mechanisms.

4.7.2 EXPOSURE

Figure 4.27 depicts the two ways that can be utilized to write waveguide structures in glass. In Figure 4.27a, the focused beam (~3 μm) is moved through the sample from the front face to the back face at a suitable speed (~0.02 mm/sec). The speed is limited by avoiding the physical damage threshold. Figure 4.27b is the more typical top-writing where the focus is below the surface and is scanned along the sample face, again at an optimum speed.

The high-order multiphoton aspect combined with the high intensity produced by the focusing objective allows the focus spot to be positioned anywhere inside the glass without producing any effect except at the focal plane. This property allows one to consider writing patterns in 3-D objects. An example is a pattern written using Fota-Lite as the medium, which was discussed in Chapter 3. This is shown in Figure 4.28.

Figure 4.29 shows patterns of waveguides written in a very wide variety of glass compositions. Figure 4.30 shows the measured modal pattern of a fs-laser-written single-mode guide along with an end view to get some idea of the size. The ability to write index changes in such a variety of glasses with widely different thermal and physical properties speaks to a mechanism independent of the specific glass composition.

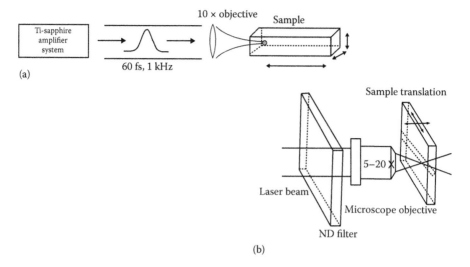

FIGURE 4.27 Schematic depiction of the two commonly used ways to write waveguide structures: (a) the focus spot is translated through the sample, and (b) the focused beam is translated from the top below the surface.

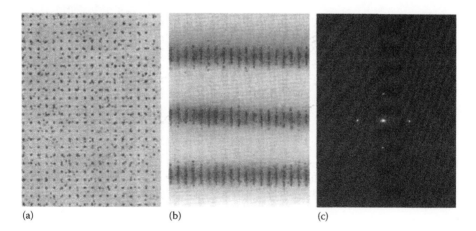

FIGURE 4.28 Images of fs-laser patterns written in photosensitive Fota-Lite glass: (a) end-on view, (b) cross section, and (c) diffraction pattern. (From N.F. Borrelli, *Microoptics Technology*, Marcel Dekker, New York, 2005.[1])

4.7.3 MECHANISM

The mechanism for refractive index change is far from being well understood. Ironically, the lack of a clear understanding of the phenomenon has not held back efforts to produce optical devices. As mentioned in Section 4.5.3, the mechanism must be considered a universal one; that is, all glasses must utilize the same process to produce the effect, although not to the same magnitude. The models that have been proposed are presented next and are abstracted from Borrelli.[1]

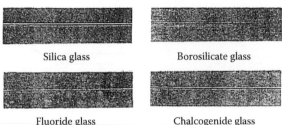

Silica glass   Borosilicate glass

Fluoride glass   Chalcogenide glass

FIGURE 4.29 Images of fs-laser waveguides written using a variety of different glasses. (Reprinted with permission from K. Miura, J. Qui, and J. Inouye, *Appl Phys. Lett.* 71(23), 3329, Copyright 1997. American Institute of Physics.[12])

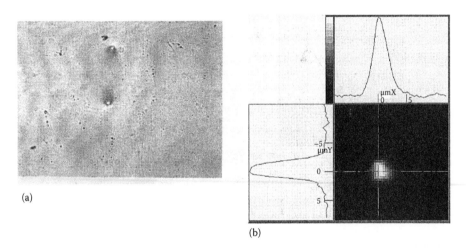

(a)

(b)

FIGURE 4.30 Photo micrograph of (a) the end view of a written waveguide, and (b) the measured modal intensity pattern. (From N.F. Borrelli, *Microoptics Technology*, Marcel Dekker, New York, 2005[1]; A. Streltsov and N.F. Borrelli, *Opt. Lett.* 26(1), 42, 2001.[14])

4.7.3.1   Color Center

The multiphoton excitation is sufficient to produce a large concentration of color centers such as SiE′ and NBOHC as will be discussed in Chapter 7, and the index change follows from the Kramers-Kronig expression. Measurements from EPR indicate that a high concentration of these centers are produced that are comparable to what can be produced by the excimer exposures mentioned in Section 4.6. However, thermal annealing experiments show that the induced index effect persists with temperature while the SiEs do not survive.[1]

4.7.3.2   Structural Change

What is involved with structural change is the realization that plasma-like electron density must be produced to initiate the structural rearrangement that leads to the refractive index change. An example of a mechanism which could involve a network bond-breaking phenomenon that could lead to a structural rearrangement is provided below.

Here, the electric field produced by the light is sufficiently large to allow an electron to tunnel to the conduction band from the valence band,

$$I(ergs/cm^2 s) = \frac{nc}{8\pi} E^2$$

For an intensity of 10^{14} W/cm^2, the field would be 2.5×10^8 V/cm. Keldysh[15] has analyzed the tunneling situation and suggests that at high intensities, tunneling is the likely mechanism whereas at the lower intensities, multiphoton absorption is the dominant way the electrons are produced. The analysis produces a parameter that is defined as

$$\gamma = \frac{\omega}{\omega_t} = \omega \frac{\sqrt{m\Delta}}{eE}$$

Here, Δ is the bandgap and m is the reduced effective mass. When $\gamma \ll 1$, the tunneling is said to be dominant, while with $\gamma \gg 1$, the multiphoton process is dominant.

The model would be that during the fs pulse the glass would undergo an electric field breakdown where ionized bonds and ions would be created and be cooled after the pulse is removed. The following rapid quench would produce a frozen structure with a different refractive index. This is consistent with the experimental fact that to anneal out the refractive index change one has to heat to above the strain point of the glass.

REFERENCES

1. N.F. Borrelli, *Microoptics Technology*, Marcel Dekker, New York, 2005.
2. K.O. Hill, Y. Fuji, D.C. Johnson, and B.S. Kawasaki, *Appl. Phys. Lett.* 32, 647, 1978.
3. G. Meltz, W.W. Morey, and W.H. Glenn, *Opt. Lett.* 14, 823, 1989.
4. P.J. Lemaire et al. *Electr. Lett.* 29, 1191, 1993.
5. R.M. Atkins et al. *Electr. Lett.* 29, 1234, 1993.
6. Wm. H. Armistead, U.S. Patent No. 4,075,024, February 1978.
7. N.F. Borrelli, G.B. Hares, and J.F. Schroder, U.S. Patent Pending.
8. C.M. Smith, N.F. Borrelli, J.J. Price, and D.C. Allan, *Appl. Phys. Lett.* 78(17), 2452, 2001.
9. J. Albert et al. *Opt. Lett.* 24(18), 1266, 1999.
10. N.F. Borrelli, D.C. Allan, and R.A. Modavis, *J. Opt. Soc. Am.* B 16(10), 1999.
11. N.F. Borrelli, C.M. Smith, and V. Bhagavatula, *Proc. SPIE* 4103, 242, 2000.
12. K. Miura, J. Qui, and J. Inouye, *Appl. Phys. Lett.* 71(23), 3329, 1997.
13. C.B. Schaffer, A. Brodeur, and E. Mazur, *Meas. Sci. Technol.* 12, 1784, 2001.
14. A. Streltsov and N.F. Borrelli, *Opt. Lett.* 26(1), 42, 2001.
15. L.V. Keldysh, *Sov. Phys. JETP* 20, 1307, 1965.

5 Photochromic Glass

To see through a glass darkly ...

Corinthians

5.1 INTRODUCTION AND BACKGROUND

Photochromic glass is likely the most familiar of all the topics in the entire book since it was and still is a commercial product of Corning Incorporated.[1,2] The trade names for the sunglass product have varied over the years with the latest being Photogray Extra™. For those who do not have or never had a pair of these sunglasses, they darken in the sunlight and fade when in the dark. A typical performance is shown in Figure 5.1.

In contrast with the majority of the other examples in the book, this photochromic property (almost) instantaneously responds to light in real time and fades when the light is removed. As a bit of a diversion, the author believes that the story of its invention is interesting enough to be told here because it is a good example of how things become commercial products (inventions) by showing more often than not how they originate from research observations made in the past that provided no inkling of its future utility. William "Bill" Armistead many years before had been working on opal glasses. Opal glasses phase-separate into two phases, one usually being a separated particulate phase. If the two phases have different refractive indices, which they most always do, then light is scattered, yielding the opal or opaque appearance. In Chapter 3, we described these types of glasses, such as Fota-Lite. In this case, Armistead was using AgCl constituent in the glass to be the initiator of the phase separation. Normally halogen ions such as Cl, I, and F are not particularly thermodynamically happy in oxide glasses so they are prone to come out of the oxide host as a separate phase, if not initially then after a thermal treatment. Opal glass products are used often for kitchenware and decorative items, among other things. Armistead had noted in his work that one of these AgCl opal glasses seemed to darken slightly when exposed to light, and more importantly, the color faded when removed from light. A number of years later when he became the director of research at the Corning Glass Works (he followed Dr. E.U. Condon of the Manhattan Project), he had visited one of Corning's ophthalmic customers (Corning made lens blanks for prescription eyeglasses) and they asked if there were any glasses that would darken in the sunlight and fade in the dark because that would make a unique sunglass. When he returned to the Corning Lab, he spoke to Don Stookey, who was the inventor of glass-ceramics, and reminded him of his observation and asked him if he could find a way to make photochromic effect in a clear glass that is a glass that would contain AgCl as the separated phase and exhibit this darkening and fading behavior that he had observed in his opal glass. Ag halide is the ingredient in photographic film that responds to light, and as we will see, much of our understanding of the darkening

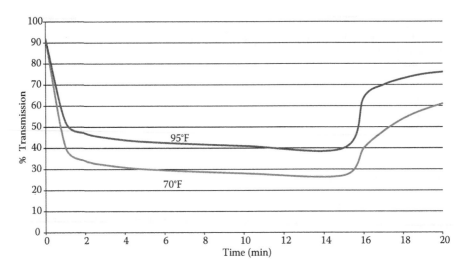

FIGURE 5.1 Representative darkening and fading performance of Ag halide-based photo-chromic sunglass.

and fading behavior of photochromic glass will appeal to the photographic model. The result through the continued genius of Don Stookey with the help of others, ultimately produced photochromic eyeglasses, although it was not an easy task, as we will see that is not only produce the effect but have properties that would make it an effective sunglass. This took a number of years to accomplish and involved the efforts, as mentioned, of many other contributing scientists. The moral of this story is to experiment with lots of things in different ways for various reasons, pay careful attention to what you see, and keep your eyes open to new possibilities. This is really what industrial laboratory research is all about.

Photochromism is known in a number of organic compounds and in a few inorganic materials.[1] A few transparent glasses have been observed to exhibit some degree of photochromic behavior. They fall into two distinct categories: homogeneous glasses such as alkali silicate glasses[2] and those such as the one to be discussed here; namely, those where a nanophase is produced from the matrix glass. In a way, one could clas-sify the glasses discussed in Chapter 7 as a photochromic effect, but we reserve the word here for those glasses that develop photochromic behavior from the formation of a second phase, which is the origin of the photochromic property. By far the biggest advantage possessed by the latter example is that recovery to clarity occurs on a time scale of minutes. There are two different metal halides that yield reversible photo-chromic behavior. The method to develop the respective nanophases as well as their physical and photochromic properties have much in common, as we will see. Initially, we will deal with Ag halide and then introduce Cu halide in Section 5.2.

The discussion in the beginning of this chapter is divided into three separate parts, beginning with Ag halide glass as the initial example that leads to the eventual production of the photochromic effect in glass. The first part deals with the glass composition space itself, since not all glasses can be made photochromic as will be shown. The second phase deals with the physics and chemistry of the production of the Ag halide phase from the glass matrix; that is, the thermodynamics of the phase separation and how it is related to the glass composition (structure). The third part deals with the physics of the photochromic property itself; that is, the physical mechanism within the Ag halide particle that determines how dark the glass gets and how fast it fades. Breaking these discussions into parts makes it easier for the reader (and also the author) to understand the phenomenon but does not imply that all the parts do not thoroughly interact to produce the end result.

5.1.1 AG HALIDE-BASED GLASS COMPOSITION

The glass composition space for any useful photochromic performance is essentially limited to the Ag halide alkali boroaluminosilicate system. This is the most representative Ag halide-based system and certainly the most studied, especially with respect to the composition-related aspects that have to do with allowing the Ag halide phase to form[3] as well as the ensuing photochromic behavior. Some discussion of other systems can be found in Araujo and Gliemmeroth and Mador[4,5] and the references found within. A sampling of the photochromic glass alkali boroaluminosilicate compositions are shown in Table 5.1. There are other important aspects of a viable glass composition having to do with providing the appropriate physical and

TABLE 5.1

Representative Compositions of Photochromic Glasses Both in Wt% and Mole%

	Wt%					
SiO_2	55.9	56.5	58.6	60.4	59.3	58.6
Al_2O_3	8.4	6.2	11.5	11.8	9.6	9.5
B_2O_3	16.2	18.2	17.5	17.7	19.7	18.3
Li_2O	2.7	1.8	2.0	2.0	2.2	1.9
Na_2O	1.9	4.1	6.7	5.9	2.1	3.0
K_2O		5.7	1.5	1.6	6.3	9.8
BaO	6.7					
ZrO_2	2.3	5.0				
TiO_2		2.1				
PbO	5.5					
	Mole%					
SiO_2	64.0	62.0	63.3	64.4	64.1	63.0
Al_2O_3	5.7	4.0	7.3	7.4	6.1	6.0
B_2O_3	16.0	17.2	16.3	16.3	18.4	17.0
Li_2O	6.2	4.0	4.3	4.3	4.8	4.1
Na_2O	2.1	4.4	7.0	6.1	2.2	3.1
K_2O		4.0	1.0	1.1	4.3	6.7
BaO	3.0					
ZrO_2	1.3	2.7				
TiO_2		1.3				
PbO	1.7		0.6	0.1	< 0.1	

TABLE 5.2

Typical Pertinent Physical Properties of Photochromic Glasses

Softening point:	663.00
Annealing point:	495.00
Strain point:	462.00
Thermal expansion ($\times 10^{-7}$):	65.00
Density:	2.41
Refractive index:	1.523
Nu value:	57.6

chemical properties that must also be satisfied, in particular those properties pertaining to the efficacy of making the glass and its ultimate stability. Table 5.2 lists the physical and thermal properties of commercial glass.

5.1.2 FORMATION OF THE AG HALIDE PHASE

This section deals with the parameters that allow Ag halide to form. The first issue is the Cl solubility; that is, how it enters the glass network, how stable it is in the melt, and how much remains in the glass as it is rapidly cooled. Equally if not more important, as discussed below, is how the Cl reacts when the glass is reheated (thermal schedule), allowing it to react with the Ag to form AgCl.

In 1948, Weyl[6] pointed out that when B_2O_3 is added to an alkali silicate in increasing amounts, many of the properties go through an extremum value. This was ascribed by Weyl to a change in the bonding of the boron as a function of composition. Bray and O'Keefe,[7] using nuclear magnetic resonance techniques, showed that in alkali borate glasses the fraction of boron atoms in tetrahedral coordination was indeed a function of composition (see Figure 5.1). Weyl pointed out further that the effect on viscosity of additions of B_2O_3 to sodium silicate glasses varies markedly with temperature. At 800°C, progressively larger additions of B_2O_3 cause monotonically decreasing values of viscosity. At 600°C, a very sharp maximum is observed. Weyl suggested that this could be understood by assuming that the bonding of boron changed with temperature as well as with composition. Araujo[8,9] attempted through the use of statistical mechanics to explain the dependence of the boron bonding on both composition and temperature. In the earlier paper,[8] he also attempted to show that bonding to trigonally coordinated boron atoms was an important mechanism for the solubility of halogens.

In glasses comprised exclusively of B_2O_3, each boron atom is bonded to three atoms of oxygen, which are in turn each bonded to two atoms of boron. All three oxygen atoms lie in a plane, and the boron is said to be trigonally coordinated. Addition of an alkali oxide modifies the bonding of the boron in any of several possible ways. When the level of alkali is low enough the addition of alkaline oxide results in boron atoms being bonded to four oxygen atoms, each of which is bonded to four boron atoms. A vicinal alkali ion compensates the negative charge, which is distributed over the four oxygen atoms bonded to the tetrahedrally coordinated boron. Addition of alkali to glasses, which contain high levels of alkali, produces an alternative modification in the bonding of the boron. The addition of one negatively charged oxygen atom produces two nonbridging oxygen atoms. In the model suggested by Araujo, halogen atoms can be bonded into the glass structure by replacing a nonbridging oxygen atom but it cannot replace a bridging oxygen atom that is bonded to a tetrahedral boron atom. Consequently, a model that allowed the calculation of the bonding of boron as a function of temperature and composition also allowed the calculation of the halogen solubility.

Abe[10] had explicitly suggested that the reason for the composition dependence of the bonding observed in alkali borates was that no two tetrahedral boron atoms could be bridged to each other through a single oxygen atom. Implicit in this suggestion was a mechanism by which the boron coordination could be a function of temperature. His suggestion implied that, at higher concentrations when there were many tetrahedral boron atoms, the inhibition of two tetrahedral boron atoms being bridged to each other placed serious constraint on the number of ways the structural units could be arranged. At low alkali levels at which not many tetrahedral boron atoms formed, the Abe rule did not result in a serious limitation on the entropy of mixing of the structural units. The fact that tetrahedrally bonded boron atoms are formed exclusively when alkali is added only in small amounts suggests that they are energetically favored.

If the bonding observed is that which produces a minimum in the free energy, the formation of nonbridging oxygens (NBOs) instead of tetrahedral borons when a modifier is added in sufficiently large amounts is readily understood. If the structure of the glass is determined by the balancing of energy and entropy, it should certainly be a function of temperature as well as composition. The fraction of boron atoms that are tetrahedrally coordinated according to the calculations described by Araujo[8,9] is shown in Figure 5.2 as a function of composition and temperature.

Having calculated values of the number of nonbridging oxygen atoms in a glass as a function of temperature and composition, Araujo calculated the solubility of chloride as a function of composition at each of two temperatures, one corresponding to the temperature at which the glass is melted and the other corresponding to the temperature at which the glass is heated to precipitate the silver halide. If the solubility exceeded some special value intended to represent the amount of chloride batched, he set the amount dissolved equal to that value. The difference in the chloride soluble (ΔCl) at these two temperatures for any given composition is taken to be the amount of chloride available for precipitation of the copper-doped silver halide phase. Very good correlation between the calculated (ΔCl) and the intensity of absorption experimentally observed at 350 nm was obtained. This absorption is believed to be due to the tail of the silver halide absorption edge. Very good correlation was also observed with the experimentally observed fade rates. The work of Yun and Bray[12] indicates that increasing levels of silica causes the straight line rise in N_4 with R to continue to larger values of R (the ratio of alkali to boron), as shown in Figure 5.3.

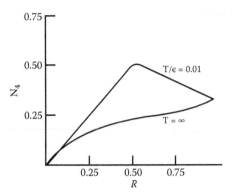

FIGURE 5.2   The upper curve is the calculated number of 4-coordinated borons as a function of the alkali to boron ratio at temperature. (From P.J. Bray and J.G. O'Keefe, *Phys. Chem. Glasses* 4, 37, 1963[7]; H. Yun and P.J. Bray, *J. Noncryst. Solids* 27, 363, 1978[12]; R.J. Araujo and J.W.H. Schreurs, *Phys. Chem. Glasses* 23, 109, 1982.[13]) The lower line is the number of 4-coordinated borons at a high temperature. (From R.J. Araujo, *J. Noncryst. Solids* 42, 209, 1980[8]; R.J. Araujo, *J. Noncryst. Solids* 58, 201, 1983.[9]) The difference between the two curves is that any alkali/boron value is taken to be proportional to the ΔCl (see text).

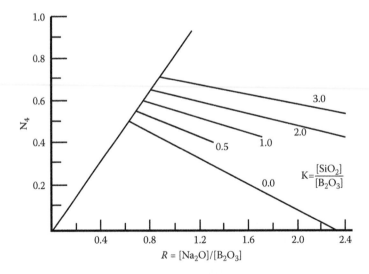

FIGURE 5.3   Number of 4-coordinated borons as a function of an alkali to boron ratio for different levels of silica to boron. (From P.J. Bray and J.G. O'Keefe, *Phys. Chem. Glasses* 4, 37, 1963[7]; H. Yun and P.J. Bray, *J. Noncryst. Solids* 27, 363, 1978.[12])

The role of alumina is considered as well; if one assumes that tetrahedral boron atoms cannot be bridged through a single oxygen atom to a tetrahedral aluminum atom, the introduction of alumina decreases the value of N_4 predicted at all but the lowest R values. This predicted effect is qualitatively in agreement with the results of Araujo and Schreurs.[13] The calculated effect of alumina on the increase in chloride,

ΔCl, seems to explain the fact that photochromism is easily obtainable over a wider range of R values in glasses that do not contain alumina than the range in glasses that do contain alumina.

5.1.3 MODEL FOR FADING

The uniquely different aspect of photochromic glasses compared to glasses that darken when exposed to light is the rapidity of its recovery after the removal from light, that is, its fading phenomenon. A model for the mechanism for fading is deemed valuable for two reasons; first, because of the physics involved, and second, because it allows one to clearly identify the compositional parameters which determine the rates that then make it possible to go back and modify either the composition or the thermal schedule in order to improve the photochromic performance.

It is generally agreed that the photochromic phenomenon results from the photo-reduction of Ag nano specks on or in the Ag halide nanoparticle. In other words, upon UV exposure the Ag halide nanocrystallites undergo a photolysis in much the same way. As well, upon exposure to ultraviolet, the silver halide crystallites in photochromic glasses undergo photolysis in much the same way as they do in photographic emulsions. Irradiation creates electron-hole pairs. The electrons are trapped, resulting in the growth of a colloidal speck of silver in or on the silver halide microcrystal.[14,15] This speck is largely responsible for the induced absorption. The major difference between the photographic emulsions and the photochromic glasses is the fate of the holes produced during irradiation. In the glasses, the hole is trapped by a Cu^+ ion in the halide microcrystal, converting it to a Cu^{+2} ion.[11,12] The reversibility of the optical absorption is due to the ability of the Cu^{+2} to recon-vert to Cu^+ with an attendant attack by its captive hole on the silver speck. In the photographic emulsion case, the hole is trapped on an uncharged halogen atom reacting with the matrix in which the Ag halide particle is suspended, and thus no reverse process is possible. Clearly, the intention in the photographic case was to have a permanent latent image that later would be developed. In the photochromic case clearly maintaining the trapped hole in the nanocrystal gives rise to the desired reversibility.

It was observed that the fading rate of silver halide-based photochromic glasses depended rather strongly on the darkening conditions. A model based on the diffu-sion of the Cu^{+2} ion to account for this behavior was proposed. In this model fading occurs if, and only if, the trapped electron (in the silver speck) and the trapped hole (Cu^{+2}) are closer than some critical distance. Since the electrons are trapped at the silver speck, they cannot wander through the silver halide crystallite and motion of the trapped hole is required to bring the hole within the critical distance from the electron. The assumption was made that the motion of the trapped hole obeyed a diffusion equation. The further assumption—that during photolysis the trapping of holes at large distances from the silver speck is improbable—is sufficient to explain qualitatively the dependence of the fade rate on the intensity of the darkening irra-diation.[13] This assumption is not unreasonable since, if an electron is ejected from a Cu^+ ion at a large distance from the silver speck, it may merely recombine with another nearby Cu^{+2} ion instead of being trapped by the silver speck.

5.1.4 QUANTITATIVE MODEL

In this section, a model for the fading is presented in an analytical format based on the thesis that the probability of recombination of an electron trapped on the silver speck and a hole trapped by a copper ion is temperature-dependent and is a continuous, monotonically decreasing function of their separation. The proposed form of the diffusion equation is given by the following.[16,17]

$$\frac{\partial \rho}{\partial t} = D\frac{\partial^2 \rho}{\partial x^2} + k_d I(A_0 - A)H(x_1) - R(x,t)$$

$$\left.\begin{array}{l}\left(\dfrac{\partial \rho}{\partial x}\right)_{x=0} = \left(\dfrac{\partial \rho}{\partial x}\right)_{x=r} = 0 \\[2mm] H(x_1) = \begin{cases} 1, & x < x_1 \\ 0, & x > x_1 \end{cases}\end{array}\right\} \qquad (5.1)$$

where D is the diffusion coefficient of the trapped hole, k_d is a constant, I is the intensity of the exciting light, x_1 is a parameter, and p represents the number of trapped holes at a distance x from the silver speck; A_0, A, and $R(x, t)$ will be discussed below. The functional form of $H(x_1)$ was chosen because it is a simple way to incorporate the assumption that electron and hole trapping is probable at small separations and improbable at large separations. An assumption was made that there is only one speck per crystallite, and that, for each trapped hole, one electron is contributed to the silver speck. The assumption that only one silver speck forms per silver halide crystallite is made to allow the mathematical simplification of being able to locate the center of a coordinate system at the silver speck. Further, if the holes are trapped very deeply, all photolytically built-up silver atoms are likely to coagulate into a single speck. (The mechanism for the formation of the Ag speck was discussed in Chapter 6.) The number of silver specks per crystallite, provided it is not very large, is not expected to change the physical results significantly. From this definition of ρ, it follows that the absorption coefficient of the glass in the darkened state can be written as the following:

$$Abs = N\sigma \int_0^r \rho\, dx \qquad (5.2)$$

where A is measured in cm^{-1}, N is the number of microcrystals of silver halide per cm^3 of glass, and σ is the absorption cross section per atom of silver that we assume to be constant. The physical significance of the parameter A_0 can now be seen to be the absorption coefficient that would be obtained if a silver atom were formed for every Cu$^+$ initially in the system. $R(x, t)$ is the spatially dependent recombination term. It is a measure of the direct recombination probability of an electron trapped in the silver speck with a cupric ion at a distance x. One assumes the recombination rate to depend on the concentration of trapped holes at a distance x and on the number of electrons available in the silver colloid. This suggests that the recombination term, $R(x, t)$, could be written by the following:

$$R(x,t) = k_r\rho(x,t)F(E)P_T(x)\int_0^r \rho(x,t)\,dx, \qquad (5.3)$$

where k is a measure of collisions per second of electrons with the surface of the silver speck, $F(E)$ is the Fermi-Dirac distribution function that takes into account the thermal occupation of the energy E, and $P_T(x)$ is the spatial dependence of the recombination process.

The tunneling model is shown in Figure 5.4. It is considered as a simple metal-insulator junction and assumes a single energy level for the cupric ion and no band bending at the interface between the silver speck and the Ag halide. In this case, we can write the following for the tunneling probability:

$$P_T(x) \propto \exp-\left[\left\{\frac{8m(V_0 - E)^{0.5}x}{h}\right\}\right] \sim \exp(-\alpha x) \qquad (5.4)$$

where h is Planck's constant, m the mass of an electron, E the energy eigenvalue, and V_0 the potential energy, in this case the conduction-band bottom.

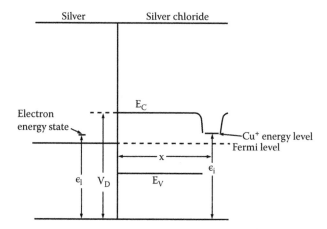

FIGURE 5.4 Interface of Ag and Ag halide treated as a metal-insulator junction with the Cu⁺ impurity level and the Fermi level shown.

Combining Equations 5.3 and 5.4, we obtain for the recombination term the following expression:

$$R(x) = \frac{k_r \rho \left\{ \int_0^r \rho \, dx \right\} \exp(-\alpha x)}{1 + \exp\left[\dfrac{E - E_F}{kT}\right]} \tag{5.5}$$

Note that the temperature dependence of $R(x)$ is solely through the thermal occupation of the energy level E in the metal by electrons. A significant temperature variation of $R(x)$ would occur only if the Cu^{+2} level were above the Fermi level of the silver halide. If one assumes that the Fermi level is in the center of the gap (1.6 eV), this is likely to be the case. Equation 5.1 is numerically integrated using the values of the constants in Table 5.3.

The value of r was taken as the diameter of the silver halide crystal determined by electron micrographs. The value of α was obtained from the WKB calculation and k_r was calculated by a transfer Hamiltonian calculation.[18] The value of $(E - E_F)$ was obtained from Moser[19] by making the assumption that the Fermi level in the silver halide was in the center of the gap; k_d was obtained by measuring the absorption coefficient of photochromic glass in the wavelength interval used for excitation and by assuming all the electrons and holes produced by photolysis are trapped. The values of A_0, x_1 and the diffusion coefficient D and its activation energy E_a were chosen so that the behavior during fading calculated for three temperatures agreed with that measured for glass ABN whose composition in wt%, which is provided in Table 5.4.

The quality of the 4-parameter fit to the fading kinetics at three different temperatures is shown in Figure 5.5.

Another interesting prediction of the model was that it predicted the experimentally observed crossover of the fading curves for the two glasses, as shown in Figure 5.6,

---

TABLE 5.3
Parameters Used for the Solution of Equation 5.1

Glossary	Parameter
k_r	10^{10} s^{-1}
α	1 Å^{-1}
ΔE	0.2 eV
x_1	15 Å
R	100 Å
A_0	50 cm^{-1}
$k_d I$	1500 trapped holes per second
$D_{Cu}^{+2} = D_0 \exp(-E_a/kT)$	$D_0 = 2.6 \times 10^{10}$ Å^2 s^{-1}
	$E_a = 0.5$ eV

TABLE 5.4

Composition of Glasses Used for Testing the Model Prediction of the Fade Rate

Glass	ABN	GBH
SiO_2	54.8	58.1
Al_2O_3	7.5	9.1
B_2O_3	29.1	17.0
Li_2O	0	1.8
Na_2O	8.1	5.6
Cs_2O	0	8.4
Ag	0.14	0.50
CuO	0.21	0.03
Cl	0.43	0.50
Br	0.26	0
CdO	0.08	0

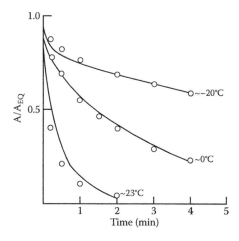

FIGURE 5.5 Numerical fit of the prediction of Equation 5.2 with the parameters listed in Table 5.3 to the experimentally measured fading curves for three temperatures. (From R.J. Araujo and N.F. Borrelli, *J. Appl. Phys.* 47(4), 1370, 1978[16]; R.J. Araujo et al., *Philos. Mag. B* 40(4), 279, 1979.[17])

at the temperatures indicated. This unexpected behavior is attributed to the interplay of the temperature dependence of the Cu^+ diffusion coefficient and the temperature dependence of the recombination term $R(x)$ in Equation 5.5; namely, $\Delta E/kT$.

The reasonable agreement of the fits gave creditability to the conjectured physical mechanism for the fading and perhaps even more importantly it points out that the significant variations in the fade rate of glasses can occur with relatively small changes in ΔE and D.[20]

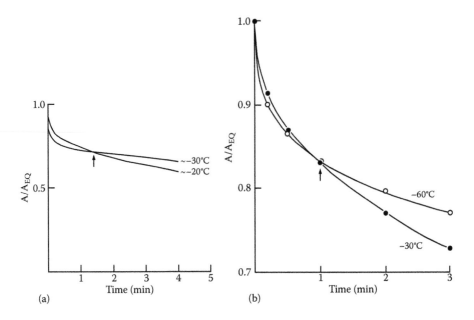

FIGURE 5.6 A model fit prediction of the observed crossover of the observed fading data at the noted temperatures. (a) Glass ABN in Table 5.4 and (b) glass GBH. (From R.J. Araujo and N.F. Borrelli, *J. Appl. Phys.* 47(4), 1370, 1978[16]; R.J. Araujo et al., *Philos. Mag. B* 40(4), 279, 1979.[17])

5.2 Cu HALIDE PHOTOCHROMIC GLASSES

As mentioned above, there is another metal halide phase that can be produced in alkali boroaluminosilicate glasses that allows reversible photochromism.[21–23] Although glasses in which a Cu halide phase can be developed share much of the same method of developing the nanophase as the Ag halide glasses, they differ quite markedly in the relation of the nanophase to the photochromic property.[23,25] Ag halide-based glasses all show photochromism to some extent, although the level of photochromism may vary when the Ag halide is present. In the Cu halide system there can be clear evidence of the existence of the Cu halide phase but exhibit no photochromism. Table 5.5 depicts the significant difference in composition of two glasses, both of which show clear evidence of the Cu halide phase, but only one is photochromic.

Referring to the previously argued value of breaking down the process into three stages—glass composition, nanophase development, and photochromic mechanism—we now see an example of perhaps some justification of that division. As we will see, although the induced absorption still is the result of photo-produced electrons being captured by metal ions leading to small metal specks in or on the halide nanocrystal, nonetheless the process by which this leads to photochromism is different.[23] An example of this is that the Cu halide photochromic glasses darken with excitation energies well below the nominal bandgap of the Cu halide. Another difference is the role the Cu halide exciton absorption feature. These differences will be discussed in the next section.

TABLE 5.5
Comparison of Cu Halide Glass Compositions Both Exhibiting the Cu Halide Phase But Only the HKX Is Photochromic

Oxide/Glass	Wt%	
	A	B
SiO_2	77.05	54.21
Al_2O_3	2.16	6.89
B_2O_3	14.16	28.75
Li_2O		
Na_2O	4.5	7.76
K_2O		
BaO		
ZrO_2		
SnO_2	0.44	0.10
CuO	0.28	1.04
WO_3		0.17
Ag		0.02
Cl^-	0.14	0.14
Br^-	0.057	0.33

5.2.1 PHOTOCHROMIC BEHAVIOR

The understanding of the photochromic behavior in the CuClBr-base system is basically split into three areas:

1. The way the glass composition (aside from the obvious indication of a Cu halide phase) and oxidizing conditions determine the presence of photochromism; not just the extent of photochromism (darkening and fading), but whether it even exists.
2. The mechanism of photochromic excitation at wavelengths less than the bandgap >400 nm (visible light).
3. The interrelationship between the photochromic property and the existence of the 400-nm exciton.

To the first point, it appears that photochromism only occurs when there is an obvious Cu^{+2} absorption feature present in the parent as-made glass. This is indicated qualitatively by the greenish tint and quantitatively by the crystal field absorption feature at 800 nm that is characteristic of Cu^{+2} in octahedral coordination. This distinguishes the photochromic compositions from the composition labeled 7351 and its variants that are water clear.[24] However, this is a necessary condition but not a sufficient one. A brief summary of compositions is provided in Table 5.5.

The following is an argument for the possible role of Cu^{+2} in producing the photochromic effect. The general picture is gleaned from the literature of what happens when a divalent ion (Ca^+ in this case) is included in an alkali halide crystal, in this example, KCl. The incorporation of the divalent ion produces a cation vacancy, as shown in Figure 5.7a.

For the cation vacancy, the argument is that the electron on Cl^- ions surrounding the vacancy are more weakly bound and thus four states above the valence band are formed. By analogy to the extent that the Cu^{+2} is electron-deficient, producing a condition similar to an anion vacancy in Figure 5.7, then an empty state below the conduction band would be created, as shown in Figure 5.7b. The conjecture is that the darkening is extended to energy less than the bandgap by a distribution of states produced by the cation vacancy caused by the Cu^{+2} substitution. A similar argument was suggested for the role of various divalent ion dopants in Ag halide. Sakuragi and Kanzaki proposed that shallow traps were produced by divalent ion substitution into the Ag halide crystal.[24]

$$K = \left(Cu_g^{+2} / Cu_g^{+1} \right) / \left(Cu_x^{+1} / Cu_x^{+2} \right) \tag{5.6}$$

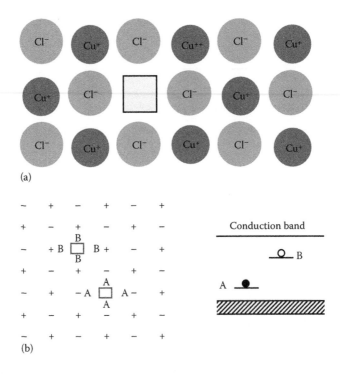

(a)

(b)

FIGURE 5.7 (a) Schematic representation of the vacancies in Cu halide crystal, and (b) the energy levels in the forbidden gap produced by the vacancy similar to that discussed by Kittel for alkali halides.

The density of cupric ions Cu^{+2} in the glass depend on (a) the level of total copper batched, (b) the concentration of other redox ions, (c) the chemical composition of the glass, and (d) the degree of oxidation of the glass. The density of cupric ions in the copper halide crystal depends on the distribution function of cupric ions between the crystal and the glass. Furthermore, the distribution function itself depends on the chemical composition of the glass. All of these affect how much Cu^{+2} ends up in the Cu halide phase. We can express this in terms of an equilibrium constant where K is a function of the specific glass composition, total Cu content, and the redox conditions of the melt.

5.2.2 EXPERIMENTAL DATA

A representative darkening and fading curve for the Cu halide photochromic glass is shown in Figure 5.7. For the figure, we see the distinct feature of ~590 nm corresponding to the Cu surface plasmon. We discussed this absorption in Chapter 2 and predicted its spectral position in Section 2.5.2. This behavior of having the excitation wavelength for darkening extend well into the visible wavelength region is quite unique in comparison to the Ag halide system discussed in the prior section. This region of darkening excitation extends into the visible wavelength region can be seen in Figure 5.8 (see also Figure 5.9).

The relative contribution from the wavelength composition of the exciting wavelength as measured as the number of transmission points of darkening is shown in Figure 5.10. A typical darkening curve at room temperature with excitation >400 nm is shown in Figure 5.11.

The darkening behavior as a function of temperature is shown in Figure 5.12a–f. As with the Ag halide discussed above, the temperature dependence has the same origin in that the recombination process (fading) is a diffusional process with attendant temperature dependence and thus directly impacts the darkened transmission

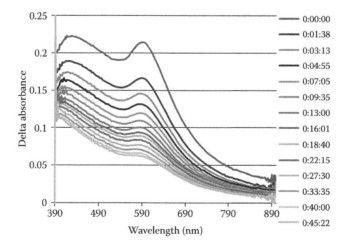

FIGURE 5.8 Spectral absorption as a function of time of fading after darkening, as noted in the legend, showing the distinct Cu exciton feature at 590 nm.

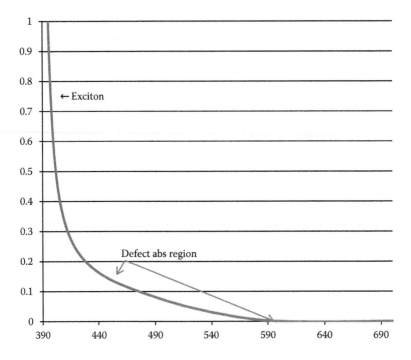

FIGURE 5.9 Absorption curve of Cu halide glass showing the spectral region (arrows) of additional absorption attributed to the Cu^{+2} defect that accounts for the visible excitation.

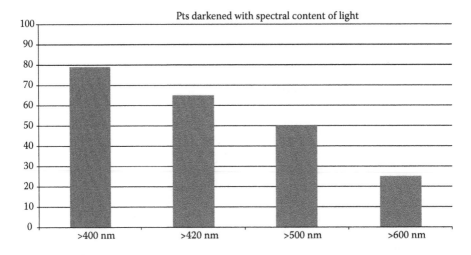

FIGURE 5.10 Comparison of the contribution to the darkening to the noted excitation wavelength regions, as noted on the graph, indicating the distribution and extent of the visible-light-induced darkening behavior.

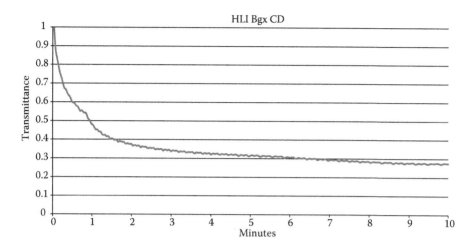

FIGURE 5.11   Typical transmission versus time for a Cu halide glass with >400-nm excitation.

level. This is quite clear in Figure 5.12f where the typical activated process is evident; the level of darkened absorption is a surrogate for the diffusion coefficient and thus the Log (abs) versus 1/T appears as a straight line with the slope indicative of the activation energy.

5.2.3   EXCITON FEATURE

The sharp cutoff absorption feature arises from the very narrow spectral width of the absorption produced by an exciton formation as compared to the absorption from a normal absorption band edge, as indicated in Figure 5.13.

The exciton absorption versus frequency is characterized by a Lorentzian band shape expressed simply as the following:

$$\text{Abs} = A / \left[\left(\omega^2 - \omega_0^2 \right) + f(\tau) \right] \tag{5.7}$$

Here, ω_0 represents the exciton resonance frequency and the $f(\tau)$ term is inversely proportional to the exciton decay time. Since the lifetime is long the line width is very narrow; hence, the rise and fall away from the resonance is very sharp. This is in vivid contrast to a normal absorption band edge that is usually described as an exponential $\exp(-a\omega)$. A comparison of the two types of absorption edge shapes is shown in Figure 5.13. Figure 5.14 shows the spectral position of the exciton for three glasses whose composition is similar to that shown in Table 5.5.

This absorption feature appears whether or not the glass shows any photochromic behavior[26]; in other words, the exciton signals the presence of the Cu halide phase but the photochromic feature depends on the specific glass composition that allows the presence of Cu^{+2}, as discussed in Section 5.2.1. In Table 5.5, the composition A is not photochromic but composition B is. This sharp absorption feature is used in

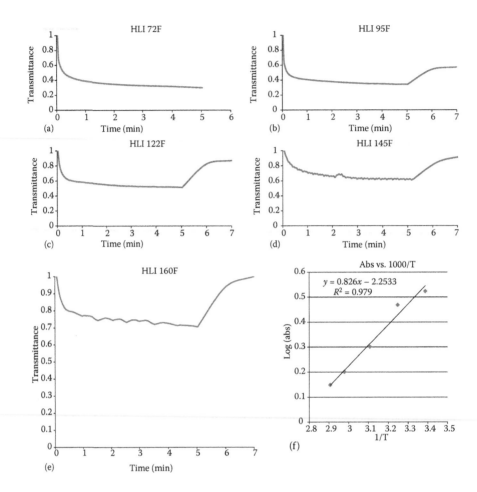

FIGURE 5.12 Temperature dependence of darkening and fading level: (a–e) with temperature noted and (f) plot of the log of the darkened absorption as a function of $T(K)^{-1}$.

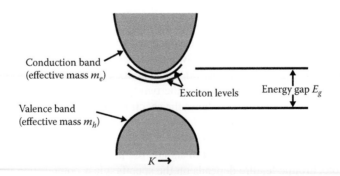

FIGURE 5.13 Exciton origin and behavior bandgap schematic showing exciton energy levels. (From D.L. Morse and T.P. Seward, U.S. Patent No. 4,222,781, 1980.[23])

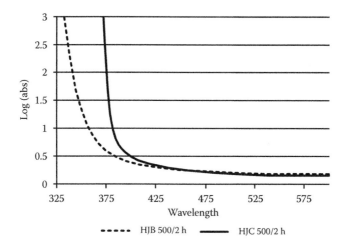

FIGURE 5.14 Absorption spectrum of glasses showing the sharp cutoff behavior in the glass resulting from the exciton of the Cu halide nanocrystal (solid curve) and a glass without the exciton (dashed curve).

an UV protecting eyeglass application when the cutoff is moved to 400 nm by adding sufficient Br to form a CuClBr phase, which has a smaller bandgap and can be tuned to 400 nm, as shown in Figure 5.14. Figure 5.15 shows where the position of the sharp cutoff property produced by the presence of the Cu halide exciton can be moved in a linear fashion by adjusting the Br/Cl ratio, while Table 5.6 shows the compositions of the glasses in the figure (see also Figure 5.16).

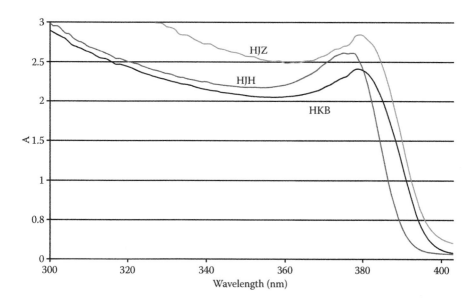

FIGURE 5.15 Absorption curves of glasses showing the spectrally resolved exciton peak.

TABLE 5.6

Compositions for the Glasses in Figure 5.15

Mol%	HJH	HJZ	HKB
SiO_2	59	59	59
Al_2O_3	4.5	4.5	4.5
B_2O_3	27	27	27
Na_2O	8.16	8.115	8.16
CuO	0.75	0.75	0.75
As_2O_3	0.25	0.25	0.25
Na_2O	0.22	0.18	0.18
Na_2O	0.12	0.14	0.14
WO_3	0	0.05	0
Ag	0	0	0.005
Co_3O_4	0	0.015	0.015
V_2O_5	0	0	0

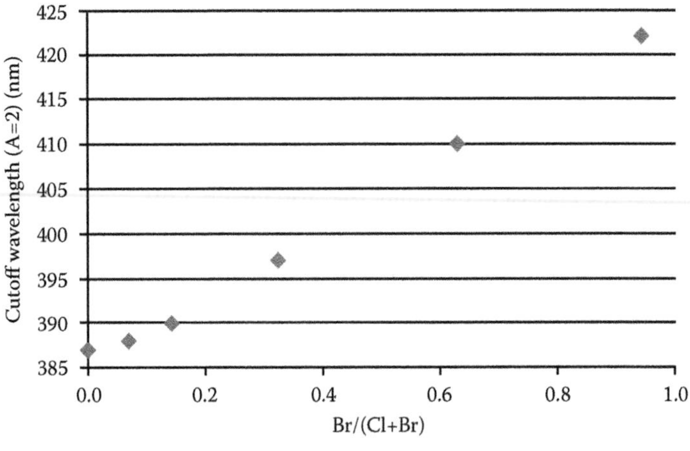

FIGURE 5.16 Position of the exciton produced an absorption cutoff of a Cu halide-containing glass as a function of the fraction of Br to total halogen.

5.3 OTHER RELATED STUDIES

In Chapter 6, a number of other interesting phenomena derived from the photochromic Ag halide phase in particular deriving from the chemistry and physics of the Ag halide nanophase will be discussed. One of the more interesting aspects involves the interaction of light with a Ag/Ag halide interface induced by UV exposure, which led to studies of Ag halide produced in nonphotochromic, which will be discussed in Chapter 7. Indeed, as the saying goes, one thing does lead to another.

REFERENCES

1. G.H. Brown, *Photochromism*, John Wiley & Sons, New York, 1971.
2. A.J. Cohen and H.L. Smith, *Science* 137, 981, 1962.
3. W.H. Armistead and S.D. Stookey, U.S. Patent No. 3,208,860, 1965; D.J. Kerko and T.P. Seward, U.S. Patent No. 4,018,965, 1977; G.B. Hares, D.L. Morse, T.P. Seward, and D.W. Smith, U.S. Patent No. 4,190,451, 1980.
4. R.J. Araujo, in *Treatise on Materials Science and Technology*, Volume 12, M. Tomozawa and R.H. Doremus (eds.), Academic Press, New York, 1977.
5. G. Gliemmeroth and K.H. Mador, *Agnew Chem. Int. Ed.* 9, 434, 1970.
6. W.A. Weyl, *The Glass Industry*, 29, 200, 1948.
7. P.J. Bray and J.G. O'Keefe, *Phys. Chem. Glasses* 4, 37, 1963.
8. R.J. Araujo, *J. Noncryst. Solids* 42, 209, 1980.
9. R.J. Araujo, *J. Noncryst. Solids* 58, 201, 1983.
10. T. Abe, *J. Am. Ceram. Soc.* 35, 284, 1952.
11. R.J. Araujo, *J. Noncryst. Solids* 71, 227, 1985.
12. H. Yun and P.J. Bray, *J. Noncryst. Solids* 27, 363, 1978.
13. R.J. Araujo and J.W.H. Schreurs, *Phys. Chem. Glasses* 23, 109, 1982.
14. C.L. Marquardt et al., *J. Appl. Phys.* 47, 4915, 1976.
15. T.P. Seward III, Third Symposium of Photochromic Glass, American Ceramic Society 76th Meeting, 1974.
16. R.J. Araujo and N.F. Borrelli, *J. Appl. Phys.* 47(4), 1370, 1978.
17. R.J. Araujo, N.F. Borrelli, and D.A. Nolan, *Philos. Mag. B* 40(4), 279, 1979.
18. C.B. Duke, *Tunneling in Solids–Solid State Physics*–Supplement 10, Academic Press, New York, 1969.
19. F.J. Moser, *Appl. Phys.* Suppl. 33, 343, 1962.
20. R.J. Araujo, N.F. Borrelli, and D.A. Nolan, *Philos. Mag. B* 50(3), 331, 1984.
21. R.J. Araujo, U.S. Patent No. 3,325,299, 1967.
22. T.P. Seward and P.A. Tick, U.S. Patent No. 3,954,485, 1976.
23. D.L. Morse and T.P. Seward, U.S. Patent No. 4,222,781, 1980.
24. S. Sakuragi and H. Kanzaki, *Phys. Rev. Lett.* 38(32), 1302, 1977.
25. V.V. Golubkov et al., *Glass Phys. Chem.* 38(3), 259, 2012.
25. C. Kittel, *Introduction to Solid State Physics*, Third Edition, John Wiley & Sons, New York, 1966.
26. A.N. Babkina et al., *Glass Phys. Chem.* 41(1), 81, 2015.

6 Photoadaption

The only thing that is constant is change.

Heraclitus

6.1 A BRIEF INTRODUCTION TO THE PHOTOADAPTION PHENOMENON

Photoadaption is a term that was adopted to describe the effect of various light exposure protocols on glasses that are considered to be nominally photochromic glass; that is, they possess a thermally developed Ag halide phase that responds to light by darkening and fading in a reversible manner, as discussed in Chapter 5. This chapter describes an additional optical phenomenon related to the photochromic effect, or more precisely, to glasses that contain a thermally developed Ag halide phase. This photoadaption was initially observed when certain photochromic glass compositions that were exposed to UV light and then exposed to light of a wavelength (subsequently or simultaneously) that was within the UV-induced absorption band (400–700 nm) resulted in a color modification.[1–5] The observed modification was that the glass took on the color of the exposure light, first seen when a He-Ne 633-nm laser was being used as probe beam to monitor the time dependence of the darkening.[1] It was observed that the probe spot became red in color. It also happened that the probe laser was linearly polarized and it was observed the red spot was also dichroic; that is, the transmission (absorption) was anisotropic with respect to the polarization direction of the light. This indicated that the optical bleaching of the absorption is more aptly described as an optical reorientation of the absorption, which will be discussed in detail in Section 5.1.

As described in Chapter 5, the photochromic effect for this special glass composition system has its origin from a silver halide nanophase that is precipitated from the glass matrix by a thermal treatment. The exposure of such glasses to ultraviolet light causes a broad absorption to develop through the visible wavelength range. This absorption is attributed to the photoreduction of silver in or on the silver halide nanocrystals by a mechanism similar to the formation of the photographic latent image, as detailed in Chapter 5. The understanding and ultimately the characterization of the morphology of the photo-produced Ag will play a major part in the explanation of the photoadaptive effects.

As we will see in later sections, photoadaption can be observed in Ag halide-containing glasses utilizing means of producing the necessary silver particle derived absorption other than UV light. Other glass compositions that manifest photoadaptive behavior can be described as additively colored. The glass composition systems where the coloration appears after cooling from the thermal treatment to develop that Ag halide phase are referred to as thermally darkenable.[5] In yet another situation, the coloration could be produced as a surface effect by adding additional Ag to the glass by ion exchange from a molten $AgNO_3$ bath after the initial thermal treatment to develop the Ag halide phase.[6,7] In a further example of the phenomenon, it was extended to a AgCl thin-film deposition technique,[8] resulting in a read-write-erasable optical memory device utilizing the changes in the polarization state that will be described.[4] In all cases a Ag metal deposit in contact with the Ag halide nanocrystals was required.

Another important aspect of understanding the photoadaptive effect has to do with the fading of the induced absorption. The discussion in Section 5.1.3 (Chapter 5) of the kinetics involved in the photochromic mechanism was only concerned with the thermal fading (thermal bleaching or fading) of the induced absorption. However, the basis of the effects discussed in this chapter is the phenomenon of optical bleaching; that is, the lessening of the induced absorption by additional exposure to light alone.[1] In Section 6.3, the proposed mechanism for the optical bleaching effect will be discussed.

In terms of the glasses that exhibit the photoadaptive phenomena, the subset of photochromic glass compositions are those that range from very slow thermal fading of the UV-induced absorption (see Sections 5.1 and 5.1.3 in Chapter 5) to those where the induced color is thermally produced. This consequently leads to glasses that develop a larger amount of induced absorption as manifested by a much deeper coloration. One key compositional difference between these photoadaptive behaviors is the much higher concentration of copper. As was pointed out in Section 5.1.4 in Chapter 5, Cu plays a major role in the photochromic process as a hole trap, but here the excess Cu seems to play the role of enhancing the amount of Ag that was produced photolytically by the UV irradiation. As mentioned above, there are other ways to introduce photoadaptive effects in these glasses other than by using UV but the composition system is basically the same irrespective of the method by which the silver absorption is produced. Representative glass compositions are shown in Tables 6.1 and 6.2 that exhibit photoadaptive behavior by various methods of production of silver absorption.

TABLE 6.1
Glass Compositions Classified as Bistable UV-Darkened

Weight%	EZC	Typical	DDT
SiO_2	52.4	61	46.3
Al_2O_3	12.9	10	1.3
B_2O_3	20.4	17	6.6
Na_2O	11.8	4.5	5.8
K_2O	13	6	0
Li_2O	0	2	2.7
Cl	0.95	0.5	6.6
Br	0.2	0	0.13
CuO	0.42	0.02	0.72
Ag	0.77	0.5	0.43
La_2O_5	0	0	0
Ta_2O_5	0	0	36.1

TABLE 6.2

Glass Compositions of Two Systems That Show the Thermally Darkened Effect

Weight%	Polarized Transmission	Wavelength
SiO_2	71	0
Al_2O_3	0	1
B_2O_3	15	25
K_2O	13	0
Li_2O	0	0
CuO	1	0
Ag	0.43	0
La_2O_5	0	50
Ta_2O_5	0	24

Source: P. Seward III, *J. Appl. Phys.* 46(2), 689, 1977.[2]

6.2 PHOTO-INDUCED COLOR AND ANISOTROPY

6.2.1 ABSORPTION BY UV EXPOSURE

The photo-induced optical anisotropy is produced in the glasses (see Table 6.1) by exposing ultraviolet-darkened glass to visible light. It is found that the ultraviolet-produced absorption band is selectively and incompletely bleached in a wavelength region corresponding to that of the bleaching light, yielding a region of color resembling the color of the bleaching light.[1-3]

If polarized light is used, the bleached region is observed to be in the sense that its absorption spectrum for light polarized parallel to that of the bleaching light is different from that of the perpendicular polarization. If unpolarized bleaching light is used, the color adaptation effect alone occurs. Figure 6.1 is an example of the time development of the induced differential transmission (dichroism), where the transmittance in both the parallel and perpendicular polarization directions are measured at 546 nm during a simultaneous exposure of the glass to ultraviolet light from a filtered Hg-arc source and polarized light is measured at 510 nm from an Ar ion laser. The inset in Figure 6.1 shows the experimental setup. Figure 6.2 shows the polarized transmission versus wavelength for two different bleaching wavelengths, indicating that the peak transmission (maximum dichroic ratio) tracks the wavelength of the bleaching light. Figure 6.3 shows how the induced dichroism expressed as a ratio of the absorption coefficients in the respective polarization directions (dichroic ratio = $\alpha_\perp/\alpha_\parallel$) varies with bleaching intensity (a) and UV intensity (b), respectively.

There are some interesting results that will have a direct bearing on the interpretation of the photoadaption mechanism. The first shows that the polarization effect is reversible and that one can convert from one dichroic orientation to another just by altering the polarization of the bleaching light, as shown in Figure 6.4.

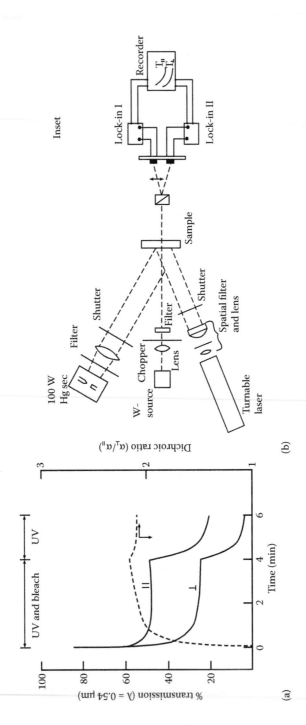

FIGURE 6.1 (a) Transmittance in both the parallel and perpendicular polarization directions as measured at 546 nm during a simultaneous exposure of the glass to ultraviolet light and polarized light at 510 nm from an Ar ion laser. (b) Diagram of the experimental arrangement. (From N.F. Borrelli, J.B. Chodak, and G.B. Hares, *J. Appl. Phys.* 50, 5978, 1979.!)

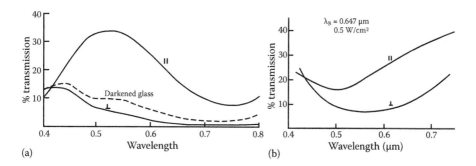

FIGURE 6.2 Polarized transmission versus wavelength for two different bleaching wavelengths: (a) 510 nm and (b) 647 nm. (From N.F. Borrelli, J.B. Chodak, and G.B. Hares, *J. Appl. Phys.* 50, 5978, 1979.[1])

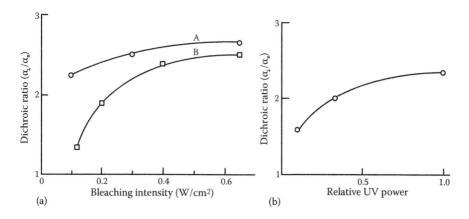

FIGURE 6.3 (a) Induced dichroic ratio as a function of the laser power at the two wavelengths noted (A = 510 nm, B = 647 nm in Figure 6.2), (b) dichroic ratio as a function of the relative UV exposure. (From N.F. Borrelli, J.B. Chodak, and G.B. Hares, *J. Appl. Phys.* 50, 5978, 1979.[1])

The glasses in which a dichroic state is induced by the polarized optical bleaching process would remain dichroic as long as they were in a darkened state. Due to the thermal fade rate of the UV-induced absorption the absorption would decrease with time (fade) and if given enough time would eventually return to virtually the clear state. A unique feature of the polarized optical bleaching phenomenon is observed when the glass is darkened with UV, then optically exposed to polarize light bleaching, and then allowed to thermally fade to the clear state in the dark. From the clear state, the glass was redarkened with UV and the polarized state returned, as shown in Figure 6.5. For many glasses the time interval between the polarizing treatment and the subsequent redarkening was as long as a year and still the dichroic property reappeared after the redarkening. This process could be repeated many times with no serious diminishment of the magnitude of the dichroic effect. It was found that heating the glasses to 200°C for a few minutes or at a lower temperature for much longer times could reduce

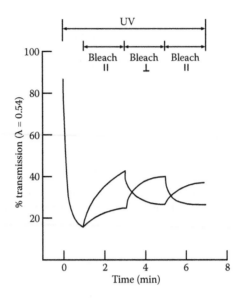

FIGURE 6.4 Graph showing how the polarization state can be reversed from one state to another by reversing the polarization direction of the bleaching source. (From N.F. Borrelli, J.B. Chodak, and G.B. Hares, *J. Appl. Phys.* 50, 5978, 1979[1]; N.F. Borrelli and T.P. Seward, *Appl. Phys. Lett.* 34, 395, 1979.[3])

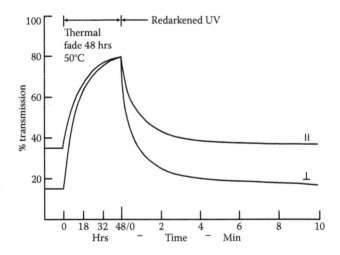

FIGURE 6.5 Graph showing the polarization state memory effect. Glass in a polarized state was left to fade to apparent clarity and then redarkened with UV light alone, and the prior polarization state is recovered. (From N.F. Borrelli, J.B. Chodak, and G.B. Hares, *J. Appl. Phys.* 50, 5978, 1979[1]; N.F. Borrelli and T.P. Seward, *Appl. Phys. Lett.* 34, 395, 1979.[3])

or eliminate the memory effect. It was also found that exposure to unpolarized light would remove the memory of the dichroic state and any induced color. We provide an explanation of these effects in a model that will be discussed in Section 6.2.4.

6.2.2 Thermally Darkened

Certain photochromic glass compositions K_2O-B_2O_3-SiO_2 La-borate systems (see Table 6.2) where the coloration is thermally developed along with the Ag halide phase, as discussed above, also produce a photoadaption.[2] Here if the level of Ag and Cu is sufficiently high, a reduction of the Ag occurs after a thermal treatment ~700°C as evidenced by a developed red color. XRD shows the presence of the Ag halide phase as a consequence of the thermal treatment although the UV-induced effect photochromic behavior is masked by the color. In Appendix 6A we will discuss the origin of the red color as resulting from the absorption of the silver in the presence of an Ag halide. One is therefore producing a similar Ag/Ag halide interface as was obtained by UV irradiation except that the color (absorption) is stronger and it does not thermally fade. With this in mind, it is not surprising that the same photoadaptive bleaching phenomena can be produced as described above for the transient darkening in photochromic glasses, as shown in Figure 6.6. In other words, the same Ag/Ag halide structure exists except that in this present case the reduced Ag is already present rather than being produced by the UV light.

The one significant advantage of this method of producing the absorbing state is that one can achieve a greater degree of induced color and dichroism; that is, the thermally reduced method produces more of the Ag/Ag halide phase than could be produced by UV exposure. This is vividly shown in Figure 6.7, where a full-colored

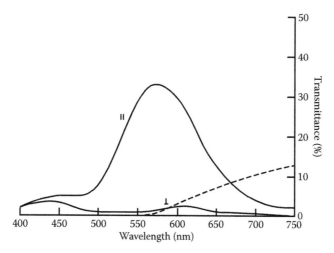

FIGURE 6.6 The curves showing the induced polarization effect (transmission in the parallel and perpendicular directions relative to the polarization of the bleaching source) glass systems in Table 6.2. The dashed line is the transmission of the glass before the polarized laser exposure. (From N.F. Borrelli, J.B. Chodak, and G.B. Hares, *J. Appl. Phys.* 50, 5978, 1979[1]; N.F. Borrelli and T.P. Seward, *Appl. Phys. Lett.* 34, 395, 1979.[3])

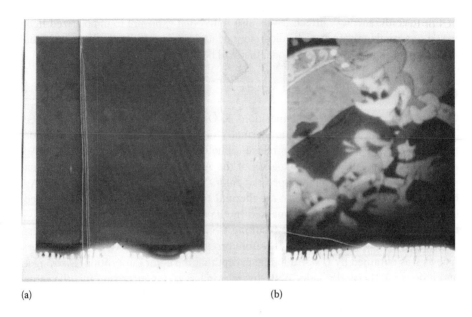

(a)   (b)

FIGURE 6.7   Demonstration of a full-color rendition of the polarizing effect by using a polarized white light source through a colored transparency: (a) viewed in unpolarized light, and (b) viewed between crossed polarizers with the sample oriented at 45 degrees.

cartoon slide was used as an exposure mask exposed with bleaching light being polarized by white light. Figure 6.7a is what is seen in unpolarized light, whereas Figure 6.7b is viewed at a 45-degree orientation with respect to crossed polarizers. With a little thought one can see how this works: Any region of the image that is not polarized will be blocked but all the polarized portions will be passed corresponding to the particular wavelength of maximum dichroic effect.

A plausible explanation for the enhanced effect in these glasses involves the size of the silver halide particles as compared the other photochromic compositions in Table 6.1. This is borne out by two observations, the first being that XRD measures more distinct peaks for this glass system. The second reason is that glasses are somewhat hazier after thermal treatment, indicating that light scattering is greater due to the larger particle size. From the larger Ag halide particle size one then asserts that the reduced silver size is proportionally larger as well. In Section 6.3, the model proposed for the origin of the photoadaption effect will make the argument as to why a larger Ag particle would show a larger effect. What is not explained is why this glass system produces a larger Ag halide crystal.

6.2.3   Ag/Ion-Exchange Method of Induced Coloration

Another related method to produce the polarizing property described above is obtained by ion-exchanging Ag into a specific class of photochromic glasses melted using an $AgNO_3$-$NaNO_3$ molten salt bath in providing the Ag in a surface layer.[6,7] The glasses, generally alkali aluminoborosilicates doped with Ag, copper, and a halogen

(Cl or Br), are then melted and formed using standard photochromic glass preparation techniques. Heat treatment was generally done at 750°C/1 h to precipitate the silver halide photochromic phase; particles of approximately 10–25-nm diameter, as confirmed by electron microscopy. The glasses were then ion-exchanged in an $AgNO_3/NaNO_3$ bath at 280°C for time ranging from 16–96 h. A deep purple color developed as a consequence of the of the Ag/ion-exchange of glasses, the degree of color depending somewhat on the glass composition and heat treatment. Bath concentration and treatment time control, primarily, the absorption magnitude and layer depth (approximately 10–150 μm). In some cases a postheat treatment was done. Bath temperature affects the induced color as does the heat treatment. Figure 6.8 shows the absorption spectra for a glass AgIX at 280°C, which was subsequently heat-treated at 400°C and 475°C, respectively. The colored layer depth also increased from 60 to 120 and 140 μm, respectively, during the thermal treatments by diffusion.

The mechanism for the IX-induced coloration involves the reduction of the additional silver supplied by the AgIX and its reduction from the Cu^{+1} in the glass.

$$Cu^{+1} + Ag^{+1} = Cu^{+2} + Ag^0 \tag{6.1}$$

Examples of exposure light sources used in the experiments are 20-mW He-Ne laser, 400-mW Krypton laser, a filtered, 150 W tungsten-halogen lamp, and a filtered, 1-KW He-Xe lamp. The lamps were generally focused to 5 cm diameter or less and the lasers to 1 cm or less. Upon exposure, the colored glass tends to take on

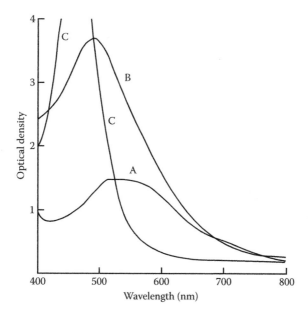

FIGURE 6.8 The initial absorption after AgIX at 280°C (C), after heating to 400°C (B), and after heating to 475°C (A). (From T.P. Seward and N.F. Borrelli, Technical Digest, Topical Meeting, "Optical phenomena in particles of small dimensions," *Opt. Soc. Am. TuB*(1–4), Tucson, AZ, 1980.[6])

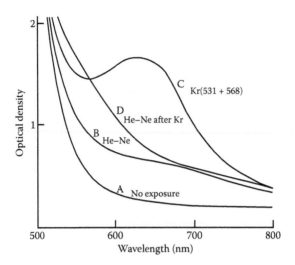

FIGURE 6.9 Absorption curves B, C and D for the AgIX glass after exposure to the indicated laser sources. The curve indicating the bleached color mimics the color of the bleaching source, and A is that for the AgIX glass before exposure. (From T.P. Seward and N.F. Borrelli, Technical Digest, Topical Meeting, "Optical phenomena in particles of small dimensions," *Opt. Soc. Am. TuB*(1–4), Tucson, AZ, 1980[6]; T.P. Seward III, *J. Non-Cryst. Sol.* 40(1–3), 499, 1980.[7])

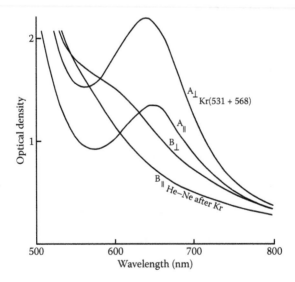

FIGURE 6.10 Induced polarizing behavior after exposure to the polarized laser sources as noted on the graph. (From T.P. Seward and N.F. Borrelli, Technical Digest, Topical Meeting, "Optical phenomena in particles of small dimensions," *Opt. Soc. Am. TuB*(1–4), Tucson, AZ, 1980[6]; T.P. Seward III, *J. Non-Cryst. Sol.* 40(1–3), 499, 1980.[7])

the color of the source; see Figure 6.9. As seen in other versions of producing the Ag/Ag halide structure (see Sections 6.2.2 and 6.2.4), a transmission region is bleached into the absorption band at the bleaching source wavelength. Further, as indicated by the example in Section 6.2.2, if the exposing source is linearly polarized, dichroism develops, as shown in Figure 6.10 where the dichroic effect is demonstrated. As discussed above, the induced dichroism is reversible in the sense that if a bleaching source of different polarization orientation is used for reexposure, the optical anisotropy develops in the new orientation. The bleached image is somewhat stable, the degree depending on glass composition. Generally the glass changes toward a less saturated color state at a rate that increases with increasing temperature.

6.2.4 THIN-FILM METHOD

Although this method does not involve glass directly, it nonetheless demonstrates the phenomenon in a more direct way and also produced an interesting practical application.[8] The photoadaptive effect discussed above was reproduced in a thin-film format by vacuum-depositing an AgCl film in conditions created to obtain color, thus emulating the Ag/Ag halide structure produced in glasses which is a necessary condition for the optical effect to be seen. In a vacuum chamber a AgCl target was used to thermally deposit a AgCl film onto a substrate.[4] Because of the reducing conditions of the deposition, some of the AgCl was reduced to Ag, imparting a purplish-red color, as shown in Figure 6.11, to the film reminiscent of the purplish-red color produced in the AgIX method described above. The film in this condition was found to exhibit the same optical bleaching phenomena that are dichroic properties when exposed to polarized visible light.

FIGURE 6.11   Glass disk coated with the deposited AgCl film.

This thin-film material possessing photoadaptive properties was investigated for a possible read-write-erasable memory. The schematic of the setup is shown in Figure 6.12. Figure 6.13 shows an optical micrograph obtained in transmission between crossed polarizers showing the visible wavelength contrast of recorded tracks; the writing energy densities were 0.75, 0.41, and 0.25 joules/cm², respectively. The tracks were recorded using a linearly polarized He-Ne laser with a 0.5 NA focusing lens with the disc traveling at a linear speed of 35 cm/sec. The writing laser was modulated with an acousto-optic device. The reading mode was with a GaAs laser (800 nm) and using a TV camera and monitor as the detector. The writing sensitivity was estimated by observing contrast at 800 nm corresponding to the

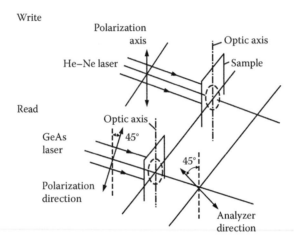

FIGURE 6.12   Schematic of the optical recording setup. (From N.F. Borrelli and P.L. Young, *SPIE 23rd Annual International Technical Symposium*, August 27–30, 1979.[8])

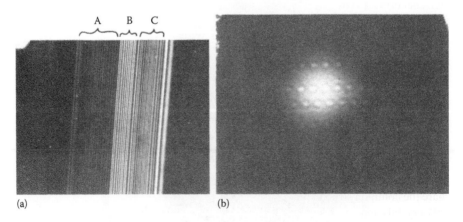

(a)   (b)

FIGURE 6.13   (a) Optical tracks of polarized bits measured through crossed polarizers Written at three recoding energies: 0.25, 0.42, and 0.75 J.cm². (b) A magnified image of a recorder bit. (From N.F. Borrelli and P.L. Young, *SPIE 23rd Annual International Technical Symposium*, August 27–30, 1979.[8])

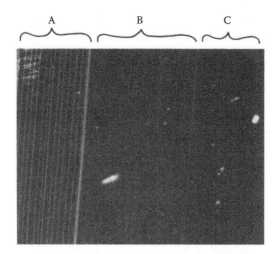

FIGURE 6.14 Erasure mode where recorded tracks were exposed to circularly polarized beam: (A) recorded tracks, (B and C) erased tracks. (From N.F. Borrelli and P.L. Young, *SPIE 23rd Annual International Technical Symposium*, August 27–30, 1979.[8])

fixed linear speed of 35 cm/sec as a function of the writing exposure. An experiment to test by exposing the tracks to circularly polarized light was done by recording a set of data tracks with writing energy density ~0.75 joules/cm^2 where the laser was circularly polarized (see Figure 6.14).

6.3   THE Ag/Ag HALIDE BLEACHING MECHANISM

The optical bleaching effect has a precedent in the photographic process called the Herschel effect.[9] Here, after a photographic emulsion is exposed, visible red light is made incident on the emulsion and results in diminished contrast after development. Mott[10] proposed a qualitative explanation of the Herschel effect in terms of a photo-excitation of an electron in the silver speck into the conduction band of the silver halide that it is in intimate contact with. The contact here is in the context of a metal insulator junction. It is proposed that this same mechanism is operative in photo-chromic glasses and is responsible for the optical bleaching effect. Further, Cameron and Taylor[11] reported a number of years ago that, if polarized light was used to bleach the photographic grains, optical anisotropy could be induced.

In this section, a physical model is proposed to explain the effects observed. The mechanism that leads to color and optical anisotropy described above would appear to be the same as is operating in the photochromic glasses reported here, since they all share the common structure of silver in contact with a silver halide phase. Unlike the bulk crystals and emulsion film, however, the optical anisotropy and adapted colors of the glasses can be erased and the effects repeated without deforming or destroy-ing the sample in photographic emulsions. If one proposes that the silver specks are anisotropically shaped (see Appendix 6B), then the degree to which Ag particles interact with the bleaching light would depend on their relative geometric orientation with respect to the polarization direction of the bleaching light. One could imagine that silver specks unfavorably oriented with respect to the polarization direction of the bleaching light will be somehow diminished in size while growth of favorably oriented specks enhanced. The color adaptation also would follow from this type of model where distribution of aspect ratios of the anisotropic silver specks are present, each absorbing most strongly at different wavelengths. Both of these explanations are explained schematically in Figure 6.15. For example, it has been shown that the peak absorption of an ellipsoidal silver particle moves toward the red wavelengths with increasing aspect ratio of the ellipsoid (see Appendix 6A). Exposure to bleaching light of a particular wavelength would diminish the population of those silver specks whose shape absorbed strongly, thus rendering the glass more transmitting at

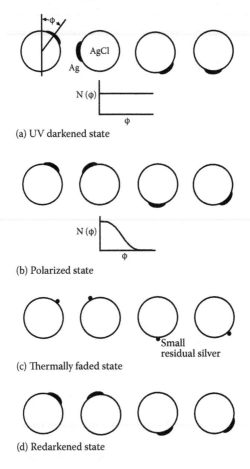

(a) UV darkened state

(b) Polarized state

(c) Thermally faded state

(d) Redarkened state

FIGURE 6.15 Schematic representation of the position of the Ag speck on the Ag halide particle: (a) initial state with Ag specks randomly formed; (b) orientation after exposure to a horizontally polarized light (From N.F. Borrelli, J.B. Chodak, and G.B. Hares, *J. Appl. Phys.* 50, 5978, 1979[1]; N.F. Borrelli and T.P. Seward, *Appl. Phys. Lett.* 34, 395, 1979.[3]); (c) residually Ag speck condition in the thermally faded condition; and (d) orientation after redarkening.

the bleaching wavelength. In fact, it has been observed that in addition to reducing the absorption at the bleaching wavelength, the absorption increases at shorter and longer wavelengths, indicating a reformation or reshaping of the silver specks during the bleaching process. A graphic demonstration of the dichroic/color-adaption effect is when these glasses are exposed to polarized white light through a multicolor image. The bleached image when viewed through polarized light shows the full-color rendition of the original image (see Figure 6.7).

Another aspect to explain is the memory effect; that is, the recovery of induced color or dichroic state after fading back to clarity. Here, one appeals again to a photographic analogy; namely, the latent image. This is the Ag speck formed in the emulsion by light exposure that is then grown by the chemical reduction step called *develop*. The time between the exposure and the subsequent development can be very long; as long as it is not exposed to light, it will remain. In the photochromic glass situation the first exposure produces the image as silver and even though the corresponding absorption can no longer be detected after fading, a latent image analog persists. This could be, for example, a deepening of the electron/hole trap discussed in Section 5.1.4 in Chapter 5 in the discussion of the photochromic effect. This then makes this position on the Ag halide more likely to trap electrons when redarkened. A schematic of the process is shown in Figure 6.15 and will be discussed in the next section.

6.4   THE MODEL

As described above, optical bleaching of *photo*chromic glasses (Ag halide) with polarizing light can lead to an optically anisotropic state that can persist after thermal fading. Although it is not clear why certain glasses can be optically bleached to a highly dichroic state while others cannot, one can make some general statements about the nature of the effect. It seems quite clear that the absorbing center in these glasses is, indeed, silver and that somehow the geometry of the silver speck that is produced photolytically gives the glass its ability to develop an optically anisotropic state via the polarized optical bleaching process. The discussion will first deal with the understanding of optical beaching and then with the origin of induced anisotropy.

6.4.1   THE OPTICAL BLEACHING MECHANISM

The optical bleaching process, that is, the mechanism by which light interacts with the silver speck in order to ultimately reduce its size, most likely proceeds by the same mechanism suggested by Mott[10] in his explanation of the Herschel effect. Essentially, it is a photoemission process whereby an electron is excited from the top of the Fermi level of the silver into the conduction band at the silver halide, as shown in Figure 6.16.[1,7]

The silver speck would be positively charged after the electron emission and would expel an Ag^+ ion from the speck to restore charge neutrality that is equivalent to a hole injection. Assuming the silver Fermi level lies somewhere near the middle of the AgCl bandgap, the threshold energy for the photoemission would be 1.6 eV

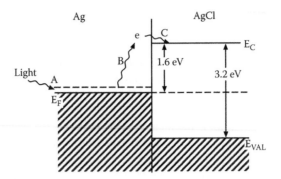

FIGURE 6.16 Ag/AgCl interface represented as a semiconductor metal insulator contact to explain the optical beaching effect. Light-induced surface plasmon absorption (A) decays by a single-electron process (B), which has sufficient absorption to promote the electron into a conduction band of AgCl (C), which then recombines with a hole that is injected into the Ag. (From N.F. Borrelli, J.B. Chodak, and G.B. Hares, *J. Appl. Phys.* 50, 5978, 1979[1]; T.P. Seward III, *J. Non-Cryst. Sol.* 40(1–3), 499, 1980.[7])

(0.77 μm). The optical bleaching efficiency, measured by the change in the absorption coefficient after bleaching divided by the absorption coefficient before bleaching as a function of the bleaching wavelength. This ratio shows a rapid decrease in the vicinity of the energy value corresponding to the Ag surface plasmon value (see Figure 6.17).

The refinement of the Mott model would be to recognize that the absorption of silver in the visible wavelength region is known to be due to free-electron plasma oscillations. The frequency of these plasma oscillations are geometry-dependent; thus, if the shape anisotropy of the optical bleaching process is to be silver-speck-geometry-dependent, then the photoemission must be produced by the plasma oscillations.

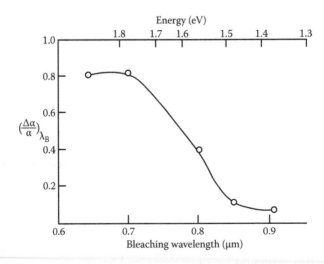

FIGURE 6.17 Efficiency of the optical bleaching as measured by $\Delta\alpha/\alpha$ as a function of energy, which shows correlation to surface plasmon absorption of the Ag nanoparticle.

Such an effect has been observed and is explained by the decay of the plasmon (collective electron excitation into a single electron excitation).[12,13]

The propensity of a given glass composition to attain a high dichroic state would then depend, by the above model, on the ability of the silver halide microcrystal to produce anisotropic specks of silver by photolysis. We have observed that the thermal heat treatment plays a significant role in determining the ultimate dichroic effect which can be produced and that likely influences the Ag halide particle size and perhaps how much Ag is thermally produced. Electron micrographs of the glass at the respective heat treatment support the particle size argument by showing a larger crystallite size for the higher-temperature heat treatments and these produce a larger dichroic ratio after the polarized light bleaching process. This suggests that the larger microcrystals allow more anisotropically shaped silver specks to form than smaller silver halide microcrystals. There is another supporting aspect to the relationship between the propensities to show the strong optical bleaching behavior with the Ag halide crystallite size. It was mentioned above that these glasses invariably have a very slow thermal fade rate. The discussion in Section 5.1.4 dealing with the kinetics of the photochromic effect itself using a diffusion model framework explicitly tied the thermal fade rate with the crystallite size; this leads to the consequence the larger the crystallite size the slower the fade rate.

6.4.2 ANISOTROPY

If the geometry of the silver speck is anisotropic, then its optical absorption will be anisotropic, and hence, its interaction with the bleaching radiation will be orientation-dependent; that is, the light will interact with the silver specks depending on their orientation with respect to the polarization direction of the bleaching light field. The net effect then would be to orient the anisotropic silver specks in space by virtue of preferentially removing particles unfavorably oriented with respect to the polarization direction to the bleaching light. If the darkening radiation is maintained with the bleaching light, then again the favorably oriented specks will grow in comparison to the unfavorable orientations. The bleaching light was incident on one edge of a rectangular sample while the transmission was measured with a beam orthogonal to this direction in the two respective polarization directions, as shown in Figure 6.18a. A UV source was maintained on the sample during the bleaching exposure. The bleaching light was unpolarized in one case, and in a second case it was linearly polarized in the direction normal to the plane formed by the bleaching beam and the monitor beam. The argument was based on the result of the reasoning of which situation would favor a particle orientation so that the major axis of the ellipsoid was in the propagation direction of the bleaching beam since it would have the smallest absorption cross section. If the particles were oblate-like, then the polarized light bleaching would favor an orientation so that the unique axis of the oblate-like particle was perpendicular to the plane containing the bleaching beam and the monitor beam. The two situations are shown in Figure 6.18b,c. The experimental result pointed to the oblate-like particle. This oblate-like shape would be consistent with the idea that the silver formed something like a thin coating on a portion of the

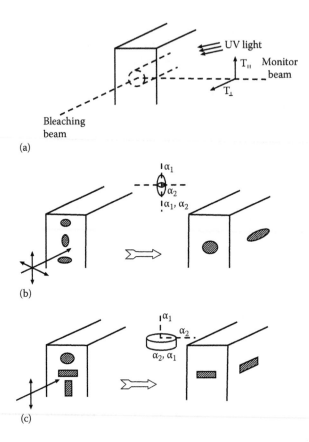

FIGURE 6.18 (a) Schematic representation of the experiment to determine the shape of the Ag speck by using the orthogonal orientation of the bleaching beam to the UV beam; (b) and (c) describe the situation for the two possible shapes of the Ag specks. (From N.F. Borrelli, J.B. Chodak, and G.B. Hares, *J. Appl. Phys.* 50, 5978, 1979.[1])

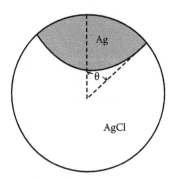

FIGURE 6.19 Schematic of the way it is proposed that the Ag deposits form on the AgCl nanoparticle. (From N.F. Borrelli, J.B. Chodak, and G.B. Hares, *J. Appl. Phys.* 50, 5978, 1979[1]; D.A. Nolan, N.F. Borrelli, and J.W.H. Schreurs, *J. Am. Ceram. Soc.* 63(5–6), 1980.[4])

surface of the spherical silver halide particle, as depicted in Figure 6.19. The coloration effect that accompanies the optical bleaching is explained in much the same manner. There is most likely a distribution of shapes and arrangements of the silver specks formed by the UV light, each by virtue of their shape and arrangement, with its characteristic absorption spectrum.

This picture is also consistent with the action of the bleaching light of a specific wavelength as it pertains to color. Here, it is to preferentially remove those specks that absorb most strongly at the bleaching wavelength. Hence, by the removal of those specks the glass becomes more transmitting at the bleaching wavelength and takes on that color. Needless to say, the process is not quite as simple in actual fact. There is evidence, as shown, that while absorption is removed at the bleaching wavelength, absorption in other wavelength regions is actually increased. This indicates that the silver formation is altered as well as becoming aligned. In general, then, one could view the process as a dynamic one where the silver deposit is being altered in response to the bleaching process, ultimately assuming a shape that is least absorbing to the bleaching wavelength.

APPENDIX 6A: THE THEORY OF SURFACE PLASMON ABSORPTION

The polarizing behavior discussed in this chapter depends on selective absorption of electromagnetic radiation by silver particles dispersed within certain oxide glasses. The absorption is due to a plasma resonance of the silver conduction electrons that is shifted into the visible or near-infrared spectral regions by the boundary conditions of the particle; specifically, the particle shape, the refractive index of the surrounding glass, and the particle orientation with respect to the applied electromagnetic radiation field.

If the particles have dimensions that are small compared to the wavelength of the radiation and are sufficiently far apart to be noninteracting, the absorption cross section per particle can be written as the following[14]:

$$C_{ABS} = \frac{2\pi V N_0^3}{L^2} \cdot \frac{B\lambda^2}{\{\varepsilon_0 - A\lambda^2 + N_0^2(1/L - 1)\}^2 + B^2\lambda^6} \tag{6A.1}$$

where V is the volume of the particle, N is the refractive index of the glass matrix, A is the wavelength in free space, and L is the electric depolarization factor[14] appropriate for the particle geometry and orientation with respect to the applied electric field. ε_0, A, and B are constants that can be approximated by calculations from the free-electron theory or obtained by fitting data for the bulk optical constants of silver. The resonant absorption occurs when the first term in the denominator is zero. The wavelength of maximum absorption, λmax, depends on the particle shape and the orientation of the polarized radiation through the electrical depolarization factor L. For particles of ellipsoidal shape, L can be determined analytically.[14] Figure 6A.1 shows how λmax varies with particle elongation for prolate and oblate shaped silver particles, respectively, in a glass, with a refractive index of 1.5, $\varepsilon_0 = 5$, and A = 55.

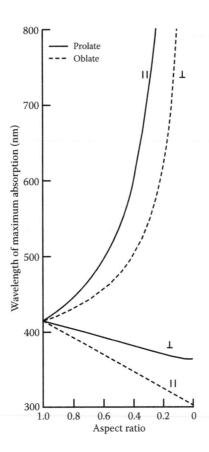

FIGURE 6A.1   Plot of the absorption maxima versus the aspect ratio of a particle for a prolate and an oblate spheroid, respectively.

APPENDIX 6B: A MODEL CALCULATION FOR INDUCED ANISOTROPY

In view of the proposed model for the induced optical anisotropy produced by polarized light bleaching of darkened photochromic glass, a relatively straightforward calculation can be made to see if the model is consistent with the magnitude of the dichroic effect observed. The model proposes that upon UV exposure, an anisotropically shaped speck of silver forms in the halide microcrystal and because of its shape, its optical absorption is anisotropic.[1] We will calculate what type of distribution in space would evolve for a given anisotropic silver shape assuming that the rate of removal of a given orientation is proportional to orientation with respect to the polarization direction of the bleaching beam.

Let $N(\Phi, \mu, \gamma)$ represent the concentration of particles of silver per cm^3 of glass oriented in space with their major symmetry axis specified by the Eulerian angles Φ, μ, and γ (see Figure 6A.2). For simplicity, we assume only one kind of silver speck

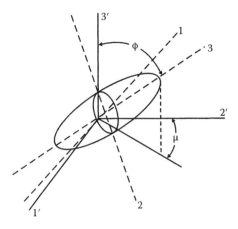

FIGURE 6A.2   Coordinate system used for the model calculation. (From N.F. Borrelli, J.B. Chodak, and G.B. Hares, *J. Appl. Phys.* 50, 5978, 1979.[1])

shape (oblate) and its aspect ratio does not change during the bleaching process. The equation governing N with time would be given by

$$dN(\varphi,\mu,\gamma)/dt = -\alpha_{\|}(\varphi,\mu,\gamma)I_{B}N + kI_{uv}\left[8\pi^2N_0 - N\int Ndv\right] \qquad (6A.2)$$

where $\alpha_{\|}$ is the absorption cross section of the speck in the polarized bleaching light direction, here taken to be at $\Phi = O$, I_{B} is the bleaching light intensity, N_0 is the concentration of Ag halide crystallites, and hence, the maximum concentration of silver specks, and kI_{uv} is proportional to the exposure intensity. Further, we assume that there is only one silver speck per particle. This model then further assumes that the action of the ultraviolet light is to produce silver specks at random orientation and is proportional only to the number of available silver halide microcrystals where no silver speck exists. The absorption cross section of the speck, referred to as its symmetry axes, can be related to the coordinate system tied to the polarization direction of the light beam by the transformation

$$\alpha'_{ij} = a_{ik}a_{jl}\,\alpha_{kl} \qquad (6A.3)$$

α' refers to the polarization direction of the bleaching light reference frame and $\alpha_{k\ell}$ to the silver speck symmetry axes, and the "a" are the direction cosines relating the two coordinate systems. Figure 6A.2 shows the coordinate system. For the sake of simplifying the calculation we assume the silver speck is an ellipsoid of revolution with α_{33} the absorption cross section for the unique axis and α_{11} the absorption cross section in the plane perpendicular to this direction. In the steady state we can solve Equation 6A.2 for the particle distribution, which is shown below for a prolate case and an oblate case, respectively.

$$N(\phi) = \frac{1}{\alpha'_{33}(\lambda_B)} \left[\frac{8\pi^2}{\beta + \frac{8\pi^2}{\sqrt{\Delta\alpha\ \alpha_{11}}} tan^{-1} \sqrt{\frac{\Delta\alpha(\lambda_B)}{\alpha_{11}(\lambda_B)}}} \right] prolate \qquad (6A.4)$$

where

$$\beta = \frac{I_B(\lambda_B)}{kI_{UV}} \Delta\alpha = \alpha_{33} - \alpha_{11}; \alpha'_{33} = \alpha_{11} + (\alpha_{33} - \alpha_{11})cos^2\phi$$

and

$$N(\phi) = \frac{1}{\alpha'_{33}(\lambda_B)} \left[\frac{8\pi^2}{\beta + \frac{8\pi^2}{\sqrt{\Delta\alpha\ \alpha_{11}}} tanh^{-1} \sqrt{\frac{\Delta\alpha(\lambda_B)}{\alpha_{11}(\lambda_B)}}} \right] oblate \qquad (6A.5)$$

where

$$\beta = \frac{I_B}{kI_{UV}}; \Delta\alpha = \alpha_{11} - \alpha_{33}; \alpha'_{33} = \alpha_{11} - (\alpha_{11} - \alpha_{33})cos^2\phi$$

Note that the distribution produced depends only on the values of the absorption cross sections of the particle at the bleaching wavelength.

Changing the bleaching wavelength will change the distribution through the wavelength dependence of absorption cross sections. With the distribution of orientations given above, the effective absorption cross section in the bleaching polarization direction and orthogonal to it, respectively, are given by

$$\left\{ \begin{array}{c} <\alpha'_{11}>_{Avg} \\ <\alpha'_{33}>_{Avg} \end{array} \right\} = \frac{1}{8}\pi^2 \int_0^{2\pi} dx \int_0^{2\pi} d\mu \int_0^{\pi} \left\{ \begin{array}{c} \alpha'_{11}(\phi) \\ \alpha'_{33}(\phi) \end{array} \right\} N(\phi) \sin\phi d\phi \qquad (6A.6)$$

The induced dichroic ratio was calculated for a particular oblate-like shape and compared to the experimental data for a given glass as shown in Figure 6A.3. In this case the bleaching wavelength was 0.647 μm. The agreement is quite good in this case; however, for other bleaching wavelengths the agreement was not as good. This is not unexpected since the model assumes one particle shape and does not take into account the change in shape during the bleaching process. Nevertheless, the intention of the simple calculation was merely to demonstrate that the correct order of magnitude of the induced dichroic effect is explainable in terms of a model where the silver speck formation on the halide microcrystal is dependent on its orientation with respect to the polarization of the bleaching light.

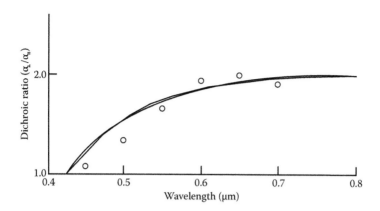

FIGURE 6A.3 Plot of the comparison of the calculated dichroic ratio versus wavelength from Equation 6A.6 with the experimental data using a 647-nm polarized bleaching source. (From N.F. Borrelli, J.B. Chodak, and G.B. Hares, *J. Appl. Phys.* 50, 5978, 1979.[1])

REFERENCES

1. N.F. Borrelli, J.B. Chodak, and G.B. Hares, *J. Appl. Phys.* 50, 5978, 1979.
2. T.P. Seward III, *J. Appl. Phys.* 46(2), 689, 1977.
3. N.F. Borrelli and T.P. Seward, *Appl. Phys. Lett.* 34, 395, 1979.
4. D.A. Nolan, N.F. Borrelli, and J.W.H. Schreurs, *J. Am. Ceram. Soc.* 63(5–6), 1980.
5. R.J. Araujo et al., U.S. Patent No. 4,125,404, 1978.
6. T.P. Seward and N.F. Borrelli, Technical Digest, Topical Meeting. "Optical phenomena in particles of small dimensions," *Opt. Soc. Am. TuB*(1–4), Tucson, AZ, 1980.
7. T.P. Seward III, *J. Non-Cryst. Sol.* 40(1–3), 499, 1980.
8. N.F. Borrelli and P.L. Young, *SPIE 23rd Annual International Technical Symposium*, August 27–30, 1979.
9. J.F.W. Herschel, *Philos. Trans. R. Soc. Lond.* 131, 1940.
10. N.F. Mott and H. Jones, *The Theory of the Properties of Metals and Alloys*, Dover Publications, New York, 1958.
11. A.E. Cameron and A.M. Taylor, *J. Opt. Soc. Am.* 24(16), 316, 1940.
12. R.A. Ferrell, *Phys. Rev.* 111(5), 1214, 1957.
13. Th. Kokkinakis and K. Alexopoulos, *Phys. Rev. Lett.* 28(2), 1632, 1972.
14. H.C. Van De Hulst, *Light Scattering by Small Particles*, John Wiley & Sons, New York, 1957.

7 Solarization

The Sun also rises ...

Ecclesiastes

7.1 BACKGROUND

The term solarization was originally used to describe the colorization of glasses produced by prolonged sunlight exposure. Early examples were purple-colored shards of glass found in the desert. In this chapter we extend the definition to include induced absorption beyond the visible portion of the spectrum. This extended definition of the solarization phenomenon has recently become of some importance in practical applications in a negative way in devices where the glass involved is exposed to UV light as, for example, from the plasma discharge in sputtering deposition units, or when an UV curing light is used to apply polymer patterns to glass. Here the understanding of the phenomenon is aimed at finding ways to mitigate the effect. It is certain that the UV portion of the spectrum (>3 eV; <400 nm) provides the relevant excitation. The reason for this range of energy of the exposure radiation is that it must be greater than the bandgap of the glass (the nature of which how it applies to glass was briefly discussed in Chapter 1). Referring to Figure 7.1, the bandgap using solid state physics terminology is the difference in energy from the electron-filled states of the valence band to the electron empty states in the conduction band. The gap is a forbidden gap, which means nominally empty of states. However, as we will see, there are other localized states in the gap. They can be multivalent impurity ions (contaminants that come in with the glass melting process) or defects that the light itself creates. These defect centers often have energy levels in the forbidden gap, as indicated in Figure 7.1. Typical centers of this sort where the energy of the light is sufficient to break a Si-O bond are the Si-E′ (silicon E-prime center) and the non-bridging oxygen hole center (NBOHC) shown in Figure 7.2.

To be able to create an electron in the conduction band and the corresponding hole in the valence band, we need sufficient energy of the photon to bridge the gap.

$$h\nu \; E_c - E_v = \text{bandgap} \tag{7.1}$$

This is the necessary first step in the process. The next step after the ionizing radiation produces electrons/holes in the conduction/valence band is the probability of the trapping of this charge[1] that would eventually recombine. The so-called traps are unoccupied energy levels near the conduction band (electron traps) and filled levels near the valence band (hole traps). It is the trapping of these species that ultimately leads directly or indirectly to the formation of the color center. These traps

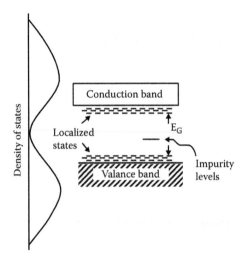

FIGURE 7.1 Schematic representation of energy levels originating from defect centers that are located within the forbidden gap.

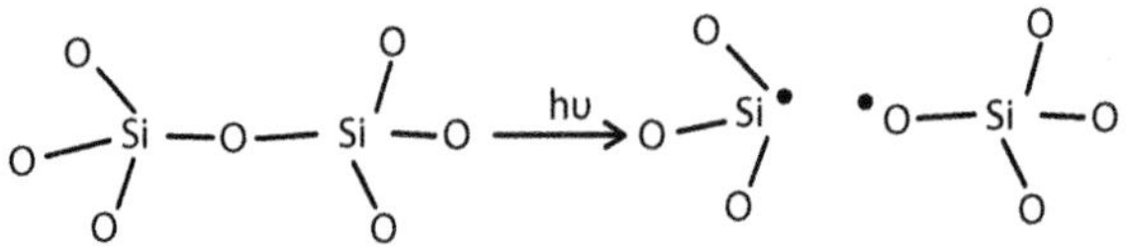

FIGURE 7.2 Schematic representation of the nonbridging hole center produced by deep UV light.

can be the defect and impurity levels mentioned above and shown in Figure 7.1. For any induced absorption to occur it is necessary that both the photo-induced electrons and holes get trapped; otherwise, rapid recombination will occur because of the high mobility of the electrons/holes in their respective bands. The efficiency of solarization that is the production of the color center is determined by the relative rates of these processes. The specific color center and therefore the position of the induced absorption depend on the glass composition as it determines the nature and number of the internal defect states, and in many cases on the impurities contained (Fe, V, Ti, etc.) in the glass.

$$h\nu \; e + h$$
$$h + T_h \rightarrow h_T \tag{7.2}$$
$$e + T_e \rightarrow e_T$$

We can write simple expressions for the rate equations conveying the general nature of the dynamic processes involved in Equation 7.3 below where e, h, represent the populations in the conduction and valence band, respectively, and e_T and h_T are the concentrations in the respective traps. T_h and T_e represent the respective empty traps and h_T and e_T represent the respective filled traps. The recombination rate constants are represented by τs (τ or recombination τ' and τ'' for electron and hole trapping, respectively, and $k(T)$ for the thermalization back into the respective bands (T is the temperature). We include for completeness possible recombination to other states nonabsorbing sites noted by τ_r:

$$\frac{de}{dt} = N_C I(v) - \frac{eh}{\tau} - \frac{e}{\tau'} - e/\tau_r$$

$$\frac{dh}{dt} = N_V I(v) - \frac{eh}{\tau} - \frac{h}{\tau''} - h/\tau'_r$$

$$\frac{de_T}{dt} = \frac{e}{\tau'} - k(T)e_T h_T$$

$$\frac{dh_T}{dt} = \frac{h}{\tau''} - k(T)'h_T e_T$$

(7.3)

The induced absorption then originates from the optical transitions associated with the respective trapped species, hole, and electron, as indicated in Equation 7.4.

$$\alpha(cm^{-1}) = \sum_i \frac{\sigma_{e_i} e_{T_i}}{\sigma_{h_i} h_{T_i}}$$

(7.4)

There is an analogy to the well-studied color centers that are optically produced in alkali halide crystals[1,2] although the physical picture is not the same. It is intended only as an example for the role of trapping sites that in the case of alkali halide crystals is attributed to ion vacancies. The case for a negative ion vacancy is shown in Figure 7.3.

Because of the diminished negative charge a photo-produced electron can be trapped. The optical transition states of these trapped electrons (called F-centers) produces the coloration. The trapped electron in this case has a small orbit and

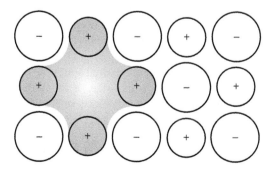

FIGURE 7.3 A depiction of a negative ion vacancy in an alkali halide crystal acting as an electron trapping site.

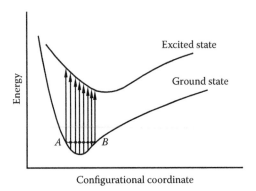

FIGURE 7.4   A schematic representation of a configuration energy diagram where energy is a function of a separation distance (phonon) leading to a broadening of the absorption feature.

thus is affected by interaction with phonons. This produces a broadened absorption feature, as would be suggested by the configurational diagram in Figure 7.4. The breadth of the induced absorption can be used as an indication of how localized the trapped species are.

7.1.1   POSSIBLE ORIGINS OF COLOR CENTERS IN GLASS

In the case of the color centers discussed above, in single crystals of alkali halides in their relation to the origin of solarization in glasses, one cannot easily define a vacancy in a similar way since the glass network is not an ordered ionic crystal nor is it likely that an oxygen or silicon atom is missing in a stoichiometric composition. However, one could consider a broader version of this situation consisting of regions in the disordered network where there is excess or depleted charge. These could be possible sites where electrons or photo-produced holes can be trapped with the trap depth that depends on the magnitude of the spatial charge imbalance at a point in the fully coordinated structure. This allows one to consider that inherent photo-induced color centers (not impurity-based) can exist through precursor atomic arrangements leading to electron/hole trapping that are ubiquitous in silicate-based glasses. It should be mentioned that electron spin resonance has been used to try to identify the trapped species.[3] In general, what one observes I resonance signals that correspond to specific ions (see Figure 7.5).

What we can conclude is that there are unpaired spin states to a level $\sim 10^{17}/cm^3$ that likely arise from a distribution of distinct sites. It should be mentioned at this point that this is not the case for light-induced defects such as SiE' and NBOHC, as we will see, where a clear resonance signal is observed for each. This will be discussed in Section 7.5.

The most reasonable place to start to investigate this phenomenon would appear to be in the ionic portion of a common silicate glass network that resides in the so-called nonbridging oxygen (NBO) produced when alkali is added to glass. Figure 7.6a–c portrays a 2-D version of a crystal, a glassy version, and one where the

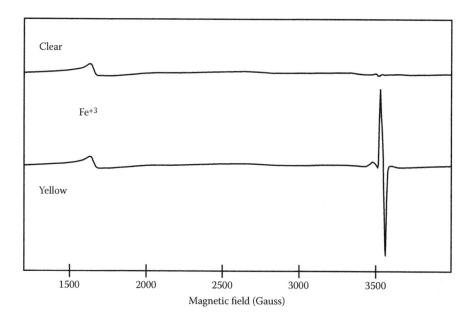

FIGURE 7.5 Electron spin resonance (EPR) spectrum of the glass containing Fe^{+3}.

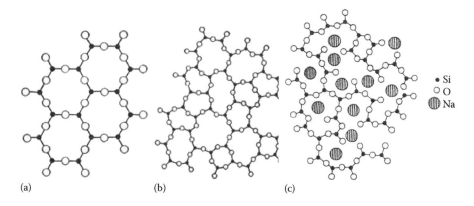

FIGURE 7.6 A 2-D depiction of (a) a crystal version of silica, (b) the disordered glassy version, and (c) an alkali containing a silica-like glass structure showing the nonbridging oxygen.

random network is interrupted to accommodate the addition of an alkali oxide to the network. The oxygen at the network break is a negatively charged oxygen ion with a positively charged alkali ion nearby to maintain charge neutrality.

Sigel[4] has demonstrated the effect of the number of NBOs on the UV edge that is related to the effective bandgap of the glass and thus determines the energy threshold for the creation of e/h pairs. This effect, expressed as the alkali concentration (one NBO for every alkali ion) is shown in Figure 7.7a. The NBO concentration as

(a)   (b)

FIGURE 7.7 A 2-D depiction of (a) an alkali borosilicate structure with no NBOs and (b) an alkali aluminosilicate with no NBOs due to the formation of 4-coordinated bonds.

a function of alkali content is not simply equal to the alkali concentration in aluminosilicates and borosilicates, respectively. In these glasses there are values of the alkali/boron and alkali/alumina ratios where there are no NBOs as a result of the fact that B and the Al are all contained in the network in 4-coordination as is the Si, as shown in Figure 7.7b.

As a consequence, there are no NBOs if the ratio of $Na_2O/Al_2O = 1$ or $Na_2O/B_2O_3 < 0.4$.[5] This is also indicated in the effect the alumina has on the bandgap, as shown in Figure 7.8.

The bandgap at the $Na_2O/Al_2O = 1$ condition appears to be close to pure silica, as one might expect from the network now being fully connected; however,

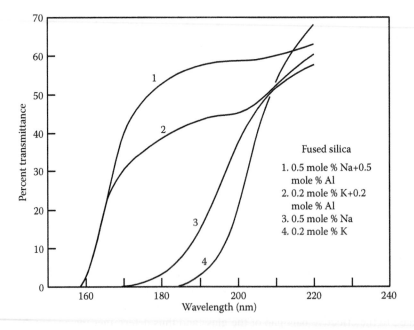

FIGURE 7.8 Shift of the UV absorption edge as a function of alumina content in fused silica. (After G.H. Sigel, *J. Phys. Chem. Sol.* 32, 2373, 1971.[4])

the electronic structure might not be completely the same. We can imagine that a collective region of boron or alumina units share the charge to compensate for the alkali ion. One related property that is strongly influenced by this configuration is the diffusion coefficient of the alkali. The solarization behavior of these systems will be discussed in more detail in the following sections. These points notwithstanding, all common silicate-based glasses seem to experience solarization to some extent, meaning they all show some induced absorption when exposed to deep UV light (<~350 nm). This means the according to the discussion above, the glasses must inherently contain possible sites where an electron/hole can be trapped and that these trapped species produce the absorption. This absorption invariably appears as a tail from the nominal before exposure band edge absorption. Electron spin resonance measures a broad spectrum of paramagnetic species. This is not to say that they all develop equivalent strength of absorption or that one cannot mitigate the magnitude or speed of the solarization by certain composition additions that would affect the trapping efficiency. The following section summarizes the experimental results for a range of glass compositions, and the reference section contains detailed solarization studies on a number of glass systems. We can easily discuss and argue almost enough to write a book dedicated to this topic alone. This chapter is in no way intended to represent a comprehensive review but rather to give the reader a basis from which to address these more detailed works. In essence, this chapter offers resources that will help to determine whether there is some common basis for the phenomenon that can be gleaned by looking at a representative set of common glasses. We have also chosen to try to broaden the phenomenon by considering the role of the electronic structure of the glass network, which so far has largely been neglected. One finds studies of specific origins and characteristics of the induced absorption, but equally important and not dealt with is the ease of electron/hole transport afforded by the glass structure. In addition, and also equally important, is the role of the electron and/or hole traps which must exist that prevent recombination of the photoionized species.

7.2 EXPERIMENTAL RESULTS

This section discusses some experimental data on the solarization phenomenon for a number of common glass systems and exposure conditions. As mentioned in the last section, this cannot be considered in any way to be a complete survey of the solarization in glass literature, which is quite extensive (see Glebov[7] and the references contained within). This discussion is intended to show a representative sampling of the different manifestations of the phenomenon so the reader can gain a better insight into the origin and physical nature of the phenomenon. More importantly is to try to understand what may be common without regard to the specific impurity.

The discussion is broken down into three parts. The first part deals with where the induced absorption clearly arises from impurities contained in the glass intentionally or otherwise; whereas the second part explores where the effect appears to arise inherently from the glass structure itself; that is, from the electronic levels of the glass. Note that the impurity effect may not be independent from the inherent effect. This means that the impurities may be involved in the overall mechanism at some point in the process even though the absorption is produced by the inherent color center. Or perhaps looking at it from the opposite view, the impurity may be the absorbing center but there is an inherent contribution from the electronic structure of the glass insofar as providing trapping centers for the electrons and holes. The third topic is rather distinct from the other two in that the color center is produced by an optical defect that is itself produced by the light. This type of defect called an SiE' center, and its complement, the NBOHC, are shown in Figure 7.2.

7.2.1 IMPURITY-INDUCED SOLARIZATION

In one sense this is the easiest case of solarization to study and understand because one has named the culprit beforehand; specifically one can if nothing else compare the degree of solarization with concentration of the named impurity. What makes it not so easy is the fact that the proposed mechanisms will more often than not vary widely depending on the specific impurity. In other words, proposed mechanisms seem ion-specific. There is a significant amount of literature devoted to this topic. As an example, more often than not these centers are paramagnetic[3] and electron paramagnetic resonance (EPR) spectroscopy can and has been extensively used to ascertain certain aspects of its local environment, allowing an interpretation of a specific electronic configuration (see Figure 7.5). This aspect notwithstanding, in general it is assumed that the photoelectrons/hole produced must travel through the respective bands at some point of the process since the UV exposure is used. Ions are not close enough for quantum mechanical tunneling to be likely. Therefore, the electrons and holes may indeed end up on specific impurity ions; the question is, how important is the way they get there? Is part of the reason for the degree of solarization to some degree dependent on the ease or lack of electron/hole transport and is this glass composition-dependent? We know for a fact that electrons and holes can travel in their respective bands; this has been shown for fused silica[6] and a sodium silicate glass.[7] The next section will deal with this in more detail. In addition, in the Appendix at the end of this chapter, unpublished photoconductivity results are presented that focus on the role of the mobility of electrons and holes on color center formation. This was measured for a number of different oxide glasses.

7.2.1.1 Ti^{+4} Impurity

This study concentrated on a series of two alkali aluminophosphate glass compositions with different levels of P and one alkali aluminoborate glass composition that all had the order of 80-ppm tramp TiO_2. The TiO_2 was unintentionally incorporated into the glass as an impurity from silica batch material. The purity of silica used as batch can be quite different depending on the source. Normally, low-iron sand is used but other impurities are still present.

Upon exposure to deep UV radiation (248-nm excimer laser, 0.1-J/pulse, 10 minutes) an absorption feature develops in the region of 550 nm, as shown in Figure 7.9a,b for alkali aluminosilicate glasses containing Ti as a contaminant. One has EPR data indicating the increase in the presence of Ti^{+3} after exposure (see Figure 7.9c) that clearly shows the presence of the Ti^{+3} signal at $g = 1.9$ and its increase with exposure

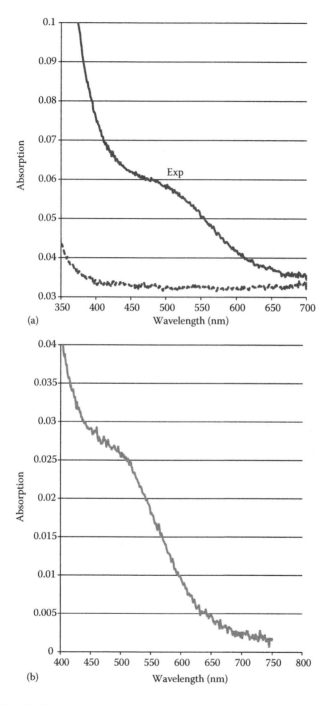

FIGURE 7.9 (a) Absorption spectrum after UV exposure of 248 nm, 2 W/cm^2, 20 Hz, 10 minutes, (b) the difference spectrum showing the induced absorption with a feature at 500 nm.   (*Continued*)

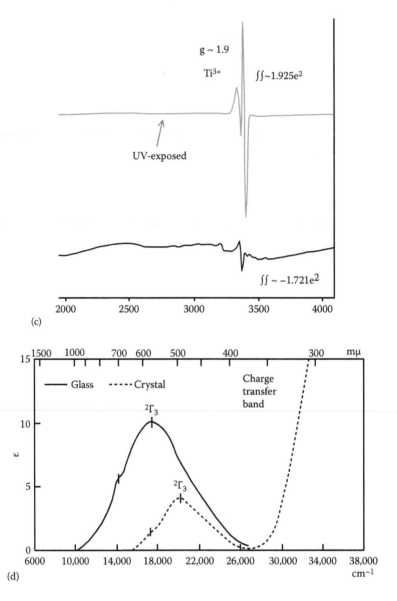

(c)

(d)

FIGURE 7.9 (CONTINUED) (c) EPR spectrum of a UV-exposed sample indicating the increase in the Ti^3 with UV exposure, and (d) absorption of Ti^{+3} in glass. (From T. Bates, in *Modern Aspects of the Vitreous State*, Vol. 2, J. D. Mackenzie (ed.), Butterworth, Washington, DC, 1962.[8])

(upper curve). The signal magnitude shows a 3× increase with UV exposure. In addition, the spectral position of the absorption peak is consistent with the crystal field transition $^2F_5 \rightarrow \, ^2F_3$ transition for Ti^{+3} in octahedral coordination depicted as the solid curve in Figure 7.9d. One can make an estimate of the induced absorption using the molar extinction coefficient from Figure 7.9d (10 cm^2/mole) and the EPR measured estimate of 400 ppm to calculate a value of 0.001/mm compared to the experimental value from Figure 7.9b of <0.01/mm.

It was further experimentally observed that the absorption could be bleached by low-intensity, deep 254-nm UV exposure, as shown in Figure 7.10. The extent of the absorption attributed to the Ti^{+3} impurities was not equal for all of the compositions for reasons not understood, as we will see in Section 7.3 when intrinsic color centers are discussed. It does appear that the induced absorption was more evident in the glasses

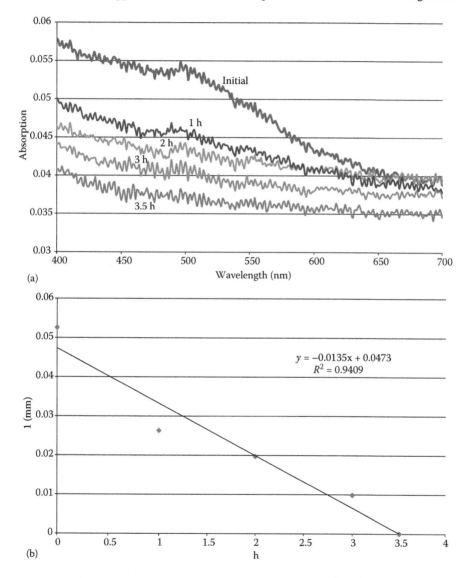

FIGURE 7.10 (a) Absorption spectra of sample with UV-induced absorption, the curve having been exposed to 254-nm light (5 mW/cm²) for 1–3.5 h showing the optical bleaching effect, (b) change in absorption coefficient at 500 nm as a function of the exposure time.

containing phosphorous although it is not clear why. One suggestion is that the Ti^{+3} with O-P ligands might have a stronger absorption cross section. This point notwithstanding, the following explanation pertains to the effect irrespective of the magnitude.

To explain the darkening and bleaching phenomena, the UV exposure wavelength must have sufficient energy to excite electrons into the conduction band. The electron then ultimately gets trapped by a Ti^{+4} to form a Ti^{+3}, which is the color center that is observed. To account for the bleaching phenomenon one must propose that the Ti^{+3} ions have excited states that lie within the conduction band. The bleaching occurs when the electron gets into the conduction band, which can then recombine with the trapped hole. This mechanism is the same that was discussed in Section 2.1 in explaining the role of Ce^{+3} as a photosensitizer for the production of Ag and Au in photosensitive glasses. In this case the electronic process involves the mixing of the d-level of the Ti^{+3} excited states with the sp^3 conduction band levels of the silicate glass. Note that the band of excited states is broad, which corresponds to the broad spectral region where bleaching occurs. Figure 7.11 schematically indicates the process. To further define the spectral range of the excitation, tripled YAG exposure (3.5 eV) produced no solarization, which suggested the range for excitation is below 300 nm (4 eV), as noted in Figure 7.11.

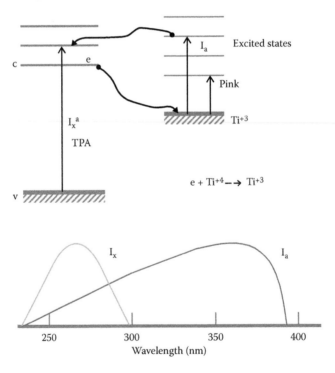

FIGURE 7.11 A schematic drawing of the bleaching process: top shows the energy diagram where the initial absorption process is indicated by the arrow labeled pink, whereas the longer arrow indicates the bleaching process where the electron is promoted back into the conduction band and recombines. The bottom drawing is an approximate sketch of the absorption cross sections of the two absorptions, excitation (darker line) and the Ti color center absorption (lighter line).

The following model is proposed: the source of the photoelectron is from the excitation from the valence band of the glass. We can account for this by first writing a rate equation for the concentration of electrons e in the conduction band:

$$\frac{de}{dt} = kIN - \frac{e}{\tau} - K(T_0 - T)e \tag{7.5}$$

where N is the electron density in the valence band and is constant, τ is the recombination, and the last term represents the trapping the electron on a Ti site, creating the Ti^{+3}. Now we can write a rate equation for the Ti^{+3} concentrations T.

$$\frac{dT}{dt} = K(T_0 - T)e - q_1 I(\lambda_1) - q_2 I(\lambda_2) \tag{7.6}$$

The respective values of qs are the rate constants for the bleaching, $I(\lambda_2) = I_b$ represents the case when there is a longer wavelength intensity component present as well.

In the steady state, we make the assumptions that $T_0 \gg T$ and $1/\tau \gg K(T_0 - T)$, which are reasonable assumptions, then the steady-state concentration of Ti^{+3} is given below:

$$T/T_0 = KkNI/\tau(q_1 I + q_2 I_b) \tag{7.7}$$

It predicts that the absorption is weakly dependent on the excitation intensity with the extent depending how large the q_2 term is. This was experimentally found to be the case. To explain the postbleaching effect with reference to Figure 7.11, we now use a source in deep UV (I_B curve) whose absorption cross section for bleaching is larger than that for darkening (I_D curve).

7.2.1.2 Other Impurity Ions

This section briefly discusses the color centers reported to be associated with other common impurities or additives such as $Fe^{+2,3}$, $Sn^{+2,3,4}$, $As^{+3,5}$, and $Sb^{+3,5}$. Iron is the dominant impurity in silicate glasses coming in with the batch materials. Because of the spectral positions of the optical absorptions stemming from the ligand field transitions of Fe^{+3} and Fe^{+2}, which is the most common impurity, both forms produced unwanted effects, the former because of the effect on the near-UV absorption leaking into the blue, and the latter because the absorption extended into the telecommunications region. Ions such as As, Sn, and Sb are also prevalent in glass not as impurities, but added as fining agents at the level of tenths of a percent by weight. A fining agent is the term used for additives that aid in the removing of gas bubbles that form in the melting of glass (primarily oxygen) because the batch materials are all invariably oxide (nitrates and carbonates). They provide this function by changing valence as a function of temperature from the melting point through the subsequent cooling by being able to transfer electrons to the oxide ion to form O_2. The O_2 bubbles then capture smaller bubbles of the other entrained gases and bring them to the surface more quickly. Mn is another common impurity that imparts color with UV exposure that we do not deal with here but its effect is shown in Section 7.4, which deals with x-ray browning. One should appreciate that solarization is the only aspect of radiation-induced color centers.

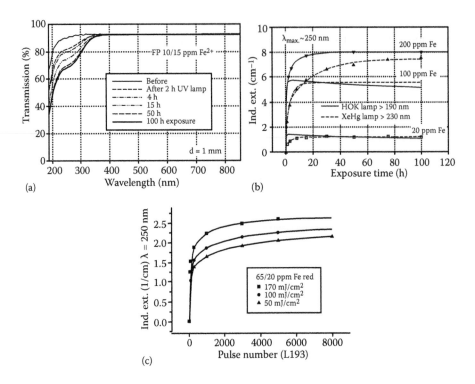

FIGURE 7.12 Spectra of Fe-activated UV induced absorption: (a) induced absorption as a function of exposure time, (b) as a function of Fe level, and (c) as a function of exposure fluence. (Reprinted from *C. R. Chim.* 5, D. Ehrt, 679, Copyright 2002, with permission from Elsevier.[9])

For Fe, the degree of solarization as a function of UV exposure of 1.5 W/cm² in the range of 230–280 nm for a fluorophosphates glass containing 10–15 ppm Fe^{+2} is shown in Figure 7.12a–c for a number of exposure protocols.[9]

The suggested mechanism for the induced absorption is attributed to a charge transfer from the Fe^{+2} to Fe^{+3} with the electron being removed to the ligand. It should be realized, however, that the process should be explained in such a way (as we have done for the Ti^{+3} and will do so below) to account for where the electron comes from (valence band) and the hole produced to be trapped on the Fe^{+2} forming a Fe^{+3}.

If indeed the photoelectron comes from the excitation of an electron from the Fe^{+2} into the conduction band, then

$$Fe^{+2} + h\nu \, Fe^{+3} + e \tag{7.8}$$

We can account for this by first writing a rate equation for the concentration of electrons e in the conduction band.

$$\frac{de}{dt} = kI(F - Fe^{+2}) - \frac{e}{\tau} - K(T_0 - T)e \tag{7.9}$$

where F is the initial Fe^{+2} concentration, τ is the recombination, and the last term represents trapping the electron on some undetermined site T. Now we can write a rate equation for the concentration for the trap T that is numerically equivalent to the produced Fe^{+3} concentration and hence the absorption.

$$\frac{dT}{dt} = K(T_0 - T)e - q_1 I - q_2 I_b \qquad (7.10)$$

The essential result is the same but the latter formulation has the advantage of allowing the hole to be transported via the valence band and the electron trapped elsewhere not necessarily localized on the ligand.

The induced absorption as produced by the excimer exposure for three levels of Sn is an alkali aluminosilicate composition (Table 7.1), which is shown in Figure 7.13a–c, and clearly shows that Sn plays a considerable role in the color center formation either as the color center itself or enhancing the effect from some other intrinsic source since there is some level of induced absorption even without Sn. EPR indicates the presence of a Sn^{3+} center in the exposed Sn-doped sample. The absorption of this center has not been identified but it seems that it is likely the source of the induced absorption. More results dealing with the effect of Sn are discussed in the next section.

Hosono[10] studied the induced absorption in As-containing soda-lime glass, his results are shown in Figure 7.14. The interpretation was based on EPR results, indicating certain possible paramagnetic As-O configurations. There is a suggestion that other defects, such as various holes, centers, and peroxy radicals, play a role. Again this makes one think that the impurity-based color centers involve other structural defects that we call inherent, which is the topic of the next section.

TABLE 7.1
Alkali Aluminosilicate Glass Compositions in Figure 7.15

Composition	A	B	C	DT
Oxide	Wt%	Wt%	Wt%	Wt%
SiO_2	66.02	57.64	64.6	60.7
Al_2O_3	13.62	21.2	13.9	12.5
B_2O_3		7.27	5.11	
Na_2O	13.73	12.78	13.75	12.1
K_2O	1.73	0.73		5.9
MgO	3.95	0.03	2.38	6.6
CaO	0.45	0.08	0.14	0.21
P_2O_5				0.08
ZrO_2				0.98
SnO_2	0.44	0.22	0.08	
SrO				0.08
R1	6.6	0.7	−0.2	8.6

Note: R1 = alkali alumina.

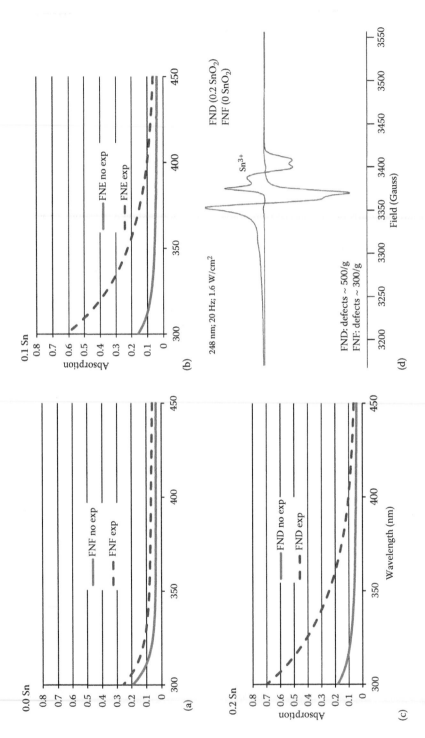

FIGURE 7.13 UV-induced absorption of alkali aluminosilicate glasses (glass A in Table 7.1): (a–c) Sn level as noted, and (d) EPR spectrum indicating the presence of an Sn^{+3} defect, exposure 248 nm, 0.6 J/pulse, 10 Hz, 10 minutes.

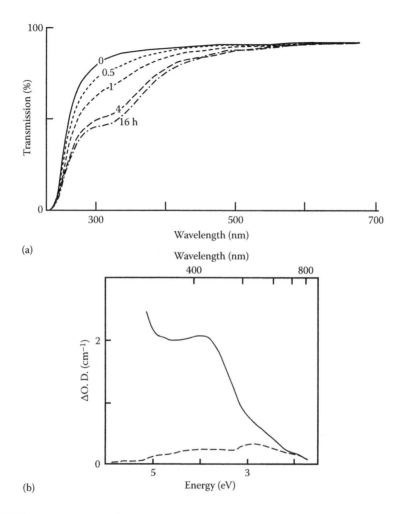

FIGURE 7.14 UV absorption attributed to As structure: (a) as a function of exposure time and (b) the induced absorption feature. (After H. Hosono and Y. Abe, *J. Non-Cryst. Sol.* 125, 98, 1990.[10])

7.3 INHERENT COLOR CENTERS

As mentioned above, it may very well be that the distinction between impurity-initiated and inherent or intrinsic-induced color center is an arbitrary one in that they share so many of the necessary conditions and only the specific manifestation of the absorption is different; that is, where in the spectrum the induced absorption occurs. Perhaps a better way to make the distinction is that the inherent part involves the role of the glass electronic structure as mentioned in Section 7.1.1, when we discussed the possible role of electron/hole transport. In other words, solarization is inherent in all glasses because—*fill in the blank*. It should be pointed out again that although all silicate glasses solarize (acquire an induced absorption when exposed to UV light), they *do not* do so to an equal extent, *and* further, the induced absorption does not necessarily appear in the same spectral region. This depends on the absorption structure of the particular color center, not to mention the strength of the absorption cross section. So, not all color centers are created equal or produce equal induced absorption effects.

As shown in Figure 7.1, oxide glasses have a region of localized states at the edge of the filled and empty bands as the result of a disordered structure. We also considered the ionic character of the structure derived from the incorporation of alkali ions and the concomitant creation of negatively charged oxygens termed nonbridging oxygens (NBOs) but the specific effect of the NBOs may differ depending on the glass structure (composition). We will deal with the latter aspect later in Sections 7.3.1 and 7.3.2. It is from these two facts and the variation in the degree of color center formation that the inherency argument must be constructed, if there is one. The other possibility of the origin of color centers is what can be called the fragments of a broken bond that are frozen in from the melt. These include things like nonbridging oxygens on the classic glass formers such as silicon (NBOHC), phosphorous (POC), and boron (BOHC) as well as the AlOHC. These centers have known EPR and optical signatures.[3]

7.3.1 ALKALI ALUMINOSILICATES

Some representative induced absorption data for alkali aluminosilicates is shown in Figure 7.15 for compositions shown in Table 7.1.

Figure 7.16a shows how the induced absorption increases with the number of 248-nm pulses and with an HgXe lamp exposure, as shown in Figure 7.16b. The impurity level in these glasses that is detailed in Table 7.1 can be considered to be approximately the same since the same batch materials were used; that is not to say that they did not all contain some degree of Fe but it would be nearly the same. They also contained different levels of Sn, as shown in Table 7.1. These differences notwithstanding, there are some differences in the induced absorption.

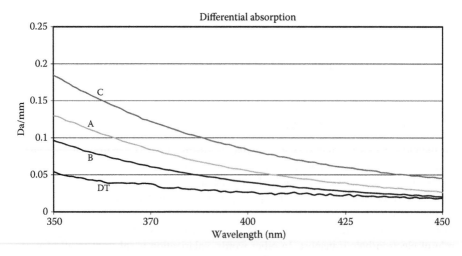

FIGURE 7.15 UV-induced absorption (∇A/mm) for four alkali aluminosilicate glasses A, B, C, and DT with compositions given in Table 7.1.

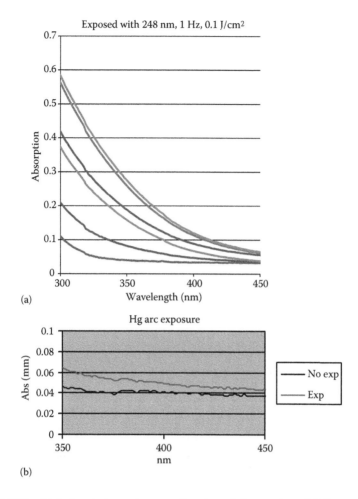

(a)

(b)

FIGURE 7.16   (a) Induced absorption as a function of the number of 248-nm exposure pulses, from bottom to top, [0, 5, 10, 50, 100] and (b) induced absorption from HgXe lamp exposure 10 mJ/cm², 10 minutes.

One other parameter that has been used to gauge photosensitivity (see Section 2.4.1) is the number of nonbridging oxygens as measured by the difference in the alkali alumina. This NBO value is reckoned from the difference in total alkali minus the alumina and is shown at the bottom of Table 7.1 as the R value. If anything, it seems to be counterindicative with the second listed composition being the most sensitive and yet having no NBOs as reckoned by the near equality of the alkali to the alumina. A further indication of the role of alumina is that shown in Figure 7.17, where the induced absorption is seen to increase markedly with the increase of the alumina content listed in Table 7.2. This data and those that follow indicate the role of the alumina in providing a greater propensity for a UV-induced color center in the visible and near-UV portion of the spectrum. This trend follows with the alkali being Li rather than Na as one observes that the induced absorption is larger for the same

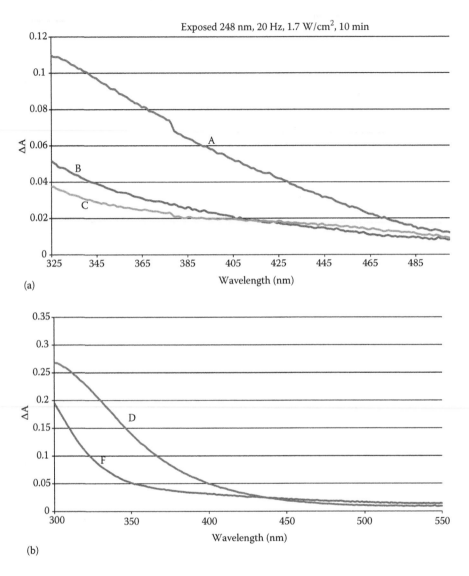

FIGURE 7.17 Induced absorption (ΔA) as a function of alumina content in an alkali aluminosilicate glass, 248 nm, 1.4 W/cm², 20 Hz, 10 min. (a) Na₂O aluminosilicate (A) 20%, (B) 15%, and (C) 12.5% alumina (see Table 7.2); (b) Li-aluminosilicate, (D) 15% and (F) 10% Al₂O₃.   *(Continued)*

alumina concentration, as shown in Figure 7.17b (see Table 7.3). One suggested candidate for the source of the induced absorption is the aluminum-oxygen hole center (AlOHC). This is the analog of the NBOHC defined above where now the bond that is broken is the Si-O-Al linkage, which would produce a SiE′ and an AlOHC.

$$-Si - O - Al + h\nu \rightarrow -Si. +.O - Al - \qquad (7.11)$$

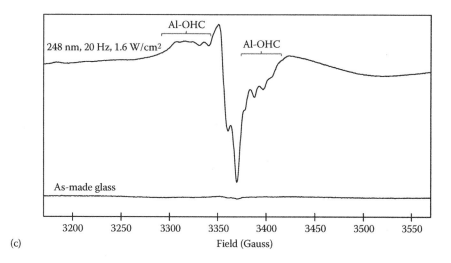

(c)

FIGURE 7.17 (CONTINUED) Induced absorption as a function of alumina content in an alkali aluminosilicate glass, 248 nm, 1.4 W/cm², 20 Hz, 10 min. (c) EPR signal of the AlOHC.

TABLE 7.2
Na-Aluminosilicate Compositions with Alumina Variation

(Mol%)	A	B	C
SiO_2	70	70	70
Al_2O_3	12.5	12.5	12.5
Na_2O	17.5	17.5	17.5

TABLE 7.3
Li-Aluminosilicate Set with Alumina Variation

(Mol%)	D	E	F
SiO_2	70	70	70
Al_2O_3	15	12.5	10
Li_2O	15	17.5	20

The ALOHC has a distinct EPR signature, as shown in Figure 7.17c. This was obtained for sample A in Table 7.2. It is noteworthy that the formation of the AlOHC was seen only in the situation when the alkali and the alumina concentrations are balanced and where the induced absorption was the largest. This happens when there are no NBOs. In the prior discussion it was proposed that the NBOs played a major role in the formation of color centers by an effective hole trap. This is not a contradiction in that indeed there is still induced absorption in the peralkaline glasses (alkali >alumina) but at a much reduced level, as seen in Figure 7.17a. The presence of the AlOHC produces a stronger color center in addition to the color produced by other mechanisms.

TABLE 7.4

Glass Compositions for the Study of the Effect of Sn on Induced Absorption

(Mol%)	A	B	C	D
SiO$_2$	70	70	70	70
Al$_2$O$_3$	12.5	12.5	12.5	12.5
Na$_2$O	17.5	17.5	17.5	17.5
SnO	0	0.22	0.44	0
SnO$_2$	0	0	0	0.2

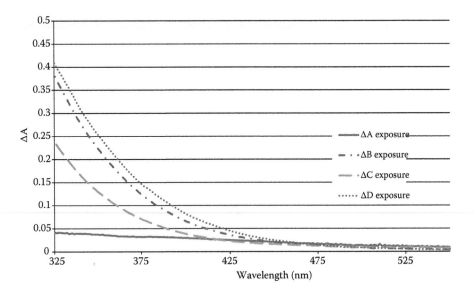

FIGURE 7.18 Induced absorption (ΔA) as a function of Sn in compositions shown in Table 7.4; 248 nm, 0.9 W/cm², 20 Hz, 10 minutes.

In Section 7.2.1.2, which dealt with the role of impurities, the effect of Sn on the induced absorption was mentioned. Here, we look at another more specific example with compositions shown in Table 7.4. Figure 7.18 shows the effect of Sn on the induced absorption. We can see a large increase in the induced absorption as a result of the addition of Sn to the glass similar to what was shown in Figure 7.13a–c.

7.3.2 ALKALI BOROSILICATES

The ternary Na$_2$O-B$_2$O$_3$-SiO$_2$ compositions as shown in Table 7.5 behave quite differently with respect to UV exposure in that the induced absorption appears less strong and is shifted much deeper into the UV portion of the spectrum relative to the aluminosilicates studied in the previous section. Moreover, more significantly they show little to no effect in the visible portion of the spectrum. These spectra are shown in Figure 7.19a,b.

TABLE 7.5

Alkali Borosilicate Composition[a]

(Mol%)	I	J	K	L
SiO_2	60	55	50	45
B_2O_3	30	27.5	25	22.5
Na_2O	10	17.5	25	32.5

[a] See text for discussion on induced absorption.

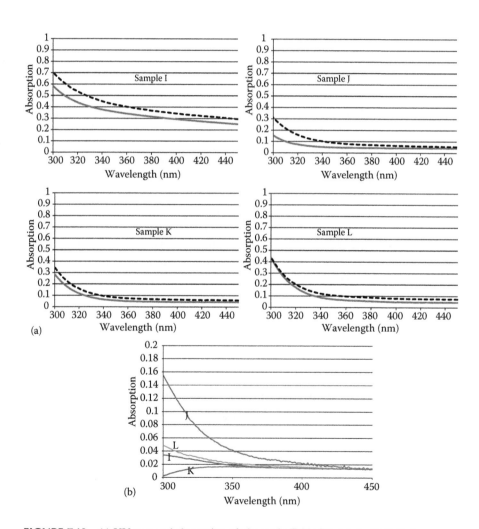

(a)

(b)

FIGURE 7.19 (a) UV-exposed absorption of glasses in Table 7.5, labeled I, J, K, L, respectively; (b) the plot of induced absorption. *(Continued)*

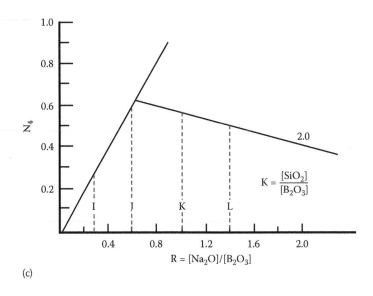

(c)

FIGURE 7.19 (CONTINUED) (c) Graph showing the relation of the boron coordination to alkali content; see text. (Adapted from H. Yun and P.J. Bray, *J. Non-Cryst. Solids* 27, 363, 1978.[5])

The first two glasses marked (I) and (J) in Figure 7.19c have alkali/boron = 0.3 and 0.63, respectively, have no NBOs; that is, all 4-coordinated borons and the latter two (K) and (L), with alkali/boron of 1 and 1.44, have some NBOs and roughly the same amount of 4-coordinated borons. This interpretation is taken from Yun and Bray and is shown in Figure 7.19c.[5] Experimental UV exposure results indicated little correlation of the degree of solarization with any of the differences in the NBO content. It appears to be the case that the sample with the most 4-coordinated borons produced the largest induced absorption.

Adding small amounts of alumina compositions, as shown in Table 7.6, produces the induced absorption difference shown in Figure 7.20.

TABLE 7.6

Compositions of Boroaluminosilicates with Increasing Values of Alumina Substitution Whose Induced Absorption Is Seen in Figure 7.20

Weight%	PAG	PAH	PAI	PAJ
SiO_2	60.88	60.88	60.88	60.88
B_2O_3	28.86	28.86	28.86	28.86
Na_2O	6.33	6.33	6.33	6.33
Na_2O	1.2	1.2	1.2	1.2
Al_2O_3	1.5	1	0.5	0.25
Sb_2O_3	0.25	0.25	0.25	0.25

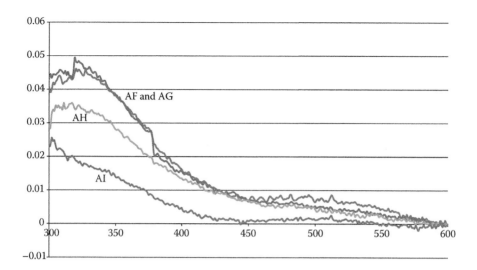

FIGURE 7.20   UV-induced absorption of alkali boroaluminosilicates shown in Table 7.7.

These glasses all should have no NBOs since the alkali exceeds the small amount of alumina added and one assumes all the boron is 4-coordinated. The induced absorption does appear to increase as the alumina concentration increases. To test this point, a set of glasses were made (see Table 7.7) where the alumina was gradually increased relative to the B_2O_3 holding the silica and alkali concentration constant.

The measured induced absorption results shown in Figure 7.21a, where one sees clearly the induced absorption increased as one increased the amount of alumina to boron concentration as seen in Figure 7.21b.

The effect of Sn on the induced absorption was also studied for one of the borosilicate glasses shown in Table 7.8. In Figure 7.22 we see little or no effect of the Sn on the induced absorption in contrast to the effect of Sn in the aluminosilicate system. This makes the role of Sn more than one of just acting as an impurity in that it acts differently depending on the glass composition.

TABLE 7.7

Alkali Borosilicates with Alumina Progressively Added in Place of Boron Oxide

(Mol%)	Q	R	S	T
SiO_2	50	50	50	50
B_2O_3	25	16.7	8.3	0
Na_2O	25	25	25	25
Al_2O_3	0	8.3	16.7	25

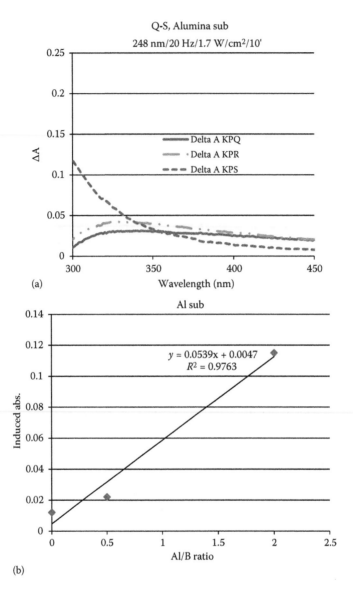

FIGURE 7.21 Induced absorption as a function of alumina content: (a) measured absorption change and (b) induced absorption versus alumina content.

7.3.3 Qualitative Mechanisms

From the above we see that the nature of the trapping levels is not known nor is how their concentration varies with the glass composition. We could simply assume that the localized levels produced as a result of the disordered network as seen in Figure 7.1 are the electron and respective electron and hole trapping sites and leave it there. It would explain why all oxide glass solarizes but would not explain why some

TABLE 7.8

Composition Set Test the Role of Sn in UV-Induced Absorption

Weight%	I	J	K	L
SiO_2	60.88	60.88	60.88	60.88
B_2O_3	28.86	28.86	28.86	28.86
Na_2O	6.33	6.33	6.33	6.33
Na_2O	1.2	1.2	1.2	1.2
Al_2O_3	2.48	0	2.48	0
Sb_2O_3	0.25	0.25	0	0
SnO	0	0	0.2	0.2

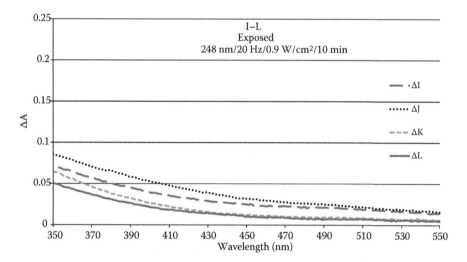

FIGURE 7.22 Induced absorption (ΔA) in Sn-containing borosilicate glasses indicating only a weak effect in contrast to the effect of Sn in the aluminosilicates.

do more than others, unless we could prove that the extent of the localization was a function of the glass composition. For example, does a less fully connected network represent a more disordered state and thus more extensive delocalization?

It appears that one way to mitigate solarization is to provide a competitive absorber in the <350-nm region. Here, we present such a result for a composition (Table 7.9) with 1200 ppm F_2O_3 added to the glass data, as shown in Figure 7.23. As discussed, the induced absorption attributed to Fe that exhibited a rather large effect that is shown in Figure 7.12b.[9] The difference is this result was obtained in a reduced glass where Fe^{+2} are dominant. For the mitigation of the induced absorption shown in Figure 7.24, the glass must be in the oxidized state having predominantly Fe^{+3}. One can easily see the difference in the UV spectrum as a consequence in Figure 7.12b where the Fe^{+3} produces the large absorption and the lack of Fe^{+2} minimizes any induced absorption from the mechanism described in the previous section.

TABLE 7.9

Composition with Added 1200 ppm Fe_2O_3 to Absorb UV to Mitigate Excitation That Produces Solarization

Weight%	AHO	AHP	AHQ	AHR
SiO_2	66.78	60.23	66.9	66.63
Al_2O_3	6.05	5.45	7.57	7.54
B_2O_3	1.55	1.4	1.55	0
Na_2O	0.92	0.83	1.84	1.83
K_2O	4.19	3.78	4.2	4.18
MgO	2.3	0	2.01	2.22
CaO	3.2	0	2.79	3.09
SrO	5.91	7.48	5.16	5.71
BaO	8.75	20.49	7.63	8.45
SnO_2	0.22	0.2	0.22	0.22
Fe_2O_3	0.12	0.11	0.12	0.12
SO_3	0.02	0.03	0.02	0.02

From all of this data it is still not clear what the actual color center is in the absence of any impurity. What is clear is that the electronic structure does play a role by providing the traps for the electron/holes that allow the color center to form. The fact that some glasses solarize more than others indicates that the trap density and depth is a function of glass composition. Further, the only way to mitigate the magnitude of the color center formation is to provide a strong competitive absorber in the wavelength region <350 nm that does not itself promote color center absorption.

However, what is significant is the more resistance to absorption in the visible wavelength region as a result of solarization compared to aluminosilicates. It is interesting to conjecture why this is so. Glass-forming oxides are oxides that are highly covalent (e.g., oxides of boron, silicon, and phosphorous). However, aluminum is much more ionic in its bonding to oxygen and consequently is not a glass-former; therefore, there are no aluminate glasses because there are silicates, borates, and phosphates. To the extent we can use the ionic crystal analogy of color center formation then the more ionic the bond the greater the propensity for charge defects, or perhaps more accurately, electronic charge fluctuations that can act as more effective electron or hole traps. This would mean that although aluminum is 4-coordinated to oxygen, the bond is somewhat closer to a nonbridging oxygen than that of a 4-coordinated boron or silicon. One needs defects with charges to produce color centers and perhaps the aluminum-oxygen bond provides this.

7.3.4 THERMAL ANNEALING

The induced absorption does thermally fade to some extent, as shown in Figure 7.25a. The activation energy for the fading seen in Figure 7.25b shows that it is a standard thermally activated process. This strongly supports the assertion that the electron

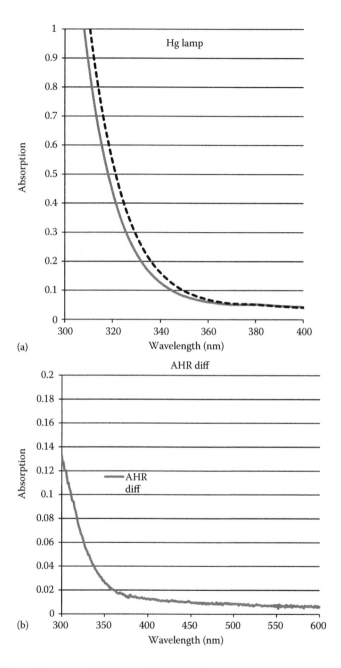

FIGURE 7.23 UV-exposed 1200 ppm F_2O_3 glass AHR shown in Table 7.4: (a) UV induced absorption (A) and (b) induced absorption (ΔA).

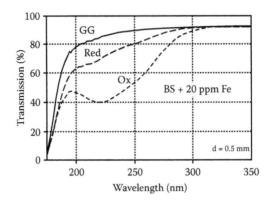

FIGURE 7.24   UV-induced absorption effect showing that it is Fe^{+3} that provided the competitive absorption. (From D. Ehrt, *C. R. Chim.* 5, 679, 2002.[9])

and holes can be thermally released into the respective bands and then can eventually recombine with the color center. This seems to confirm the conjecture that the electronic bands are involved with the process. In other words, the trapped species can be thermalized into the respective band and then recombine. There has been no evidence of optical bleaching involving inherent color centers, which is in contrast to what we have shown for impurity-based color centers.

7.4   X-RAY-INDUCED SOLARIZATION

The use of x-rays can considerably enhance the solarization phenomenon because of the much higher number of electron/hole pairs that are produced. An estimate of the enhancement of the exposure is obtained by dividing the x-ray energy (60 keV in this case) by the bandgap energy ~5 eV, which is roughly 10^4 times more. This effect was seen in the old CRT monitors where an image could be permanently burned into the screen if it remained there for a long period of time. This radiation browning, as it was called, was from the x-rays produced from the scanning electron beam incident on the glass tube face. The problem was eventually lessened by a glass composition modification. This system eventually evolved into the screen saver of today.

Figure 7.26a shows some representative colors produced by x-ray exposures as a result of adding small amounts of impurities such as Fe, Mn, and Ti to the glass.[11] The absorption spectra are shown in Figure 7.26b. Representative glass compositions shown are similar to those in Table 7.10.

One advantage that results from x-ray exposure is better spectral definition of the color center. The mechanism of the effect is essentially the same as that discussed above from UV exposure, both the browning from the inherent effect and the colors produced by the impurities. One difference is that the x-ray radiation likely created additional defects that trap the holes or electrons, thus contributing to the overall enhancement.

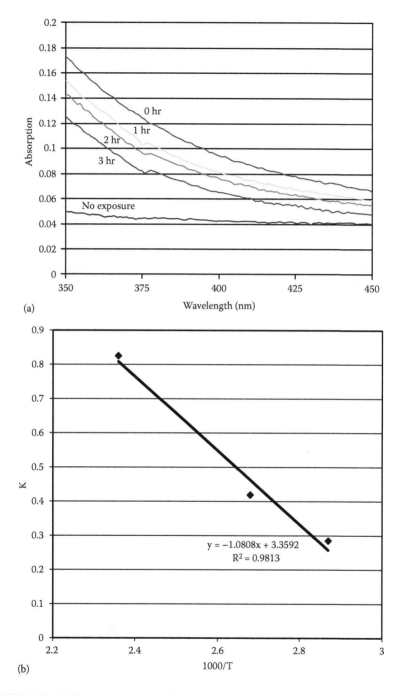

(a)

(b)

FIGURE 7.25 (a) Thermal bleaching of induced absorption versus time. (b) Absorption coefficient versus 1/T appears to be an example of a thermally activated mechanism.

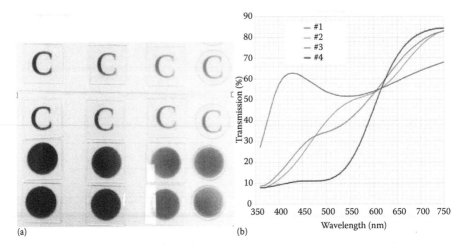

(a)   (b)

FIGURE 7.26 (a) Colors produced by exposure to 60-keV x-rays in a number of alkali alu-minosilicate glasses, and (b) representative transmission curves of the exposed glasses from Table 7.6.

TABLE 7.10
Representative Glass Compositions 1–4 That Were X-Ray-Colored

Glass Composition	#1	#2	#3	#4
SiO_2	67.734	68.210	57.854	58.398
Al_2O_3	11.084	12.587	14.678	14.593
B_2O_3	1.339	1.887	0.027	0.028
P_2O_5	0.000	0.000	7.595	7.710
Na_2O	12.357	10.737	14.947	14.790
K_2O	0.828	0.692	0.500	0.476
MgO	5.611	5.244	1.466	0.426
CaO	0.683	0.307	0.034	0.051
SnO_2	0.101	0.088	0.068	0.070
MnO_2	0.205	0.187	0.002	0.001
ZrO_2	0.016	0.022	0.042	0.044
TiO_2	0.008	0.008	2.782	3.409
Fe_2O_3	0.026	0.025	0.006	0.004
SrO	0.006	0.006	0.000	0.000
Color	Purple	Purple	Brown	Brown
Exposed color			Purple	Purple

7.5 LIGHT-INDUCED DEFECTS

The last induced color center to be discussed is unique in that the wavelength and the intensity of the excitation light is sufficient to produce a structural defect and correspondingly the color centers associated with it.[12,13] The production of this type of color is produced in a range of silicate and germanosilicate glasses[14] however the study in fused silica has been the most extensive and displays the phenomenon in a comprehensive manner so we will primarily deal with it in this section. Although E′ and NBOHC (see Figure 7.27) are produced in multicomponent silicate to a lesser extent than in silica, it is assumed to be the same mechanism as in fused silica.[15,16]

Another system that is well studied in terms of the effects produced by deep UV exposure is the GeO_2-SiO_2[14] for the induced refractive index effect (photorefraction) property, which was covered in Chapter 4. The induced refractive effect produced by deep UV exposure is certainly related to the induced absorption through the Kramers-Kronig relation; however, this is not a mechanism but a result. In other words, induced absorption will always lead to a refractive index change with wavelength in the manner described by the K-K expression: the larger the magnitude of the induced absorption, the larger the index change. In the type of color centers discussed in Sections 7.2 and 7.3, the induced absorption is relatively small compared to type of centers described here. The difference depends on two major differences, the first being the concentration of defects that can be produced, and the second the magnitude of the absorption cross section of the particular transition involved. In the impurity case discussed in Section 7.2, it is clearly limited by the impurity concentration, which is always quite small. The concentration sites for the inherent defects is not really known and for the case of the direct produced color center discussed here, it could be quite high if every Si-O bond is a possible site.

Aside from the scientific interest in color center formation is concerned an important practical relevance to the nature of the direct color center formation arises in the use of fused silica to fabricate lens elements in a high-resolution photolithographic system where excimer laser irradiation at either 248 or 193 nm is used. Irradiation at either wavelength causes induced absorption and of course because of the prolonged exposure over the lifetime of the optical element (billions of pulses in the $\mu J/cm^2$ range), it is important to understand the origin of the effect and perhaps with that understanding mitigate the induced absorption in some significant way. It is seen that deep UV irradiation of silica gives rise to induced absorption exhibiting a distinct peak at 215 nm. This absorption peak in fused silica has been ascribed to the presence of SiE′ centers.[15] This peak is broad enough so that it contributes to absorption at 193 nm and the total absorption at 193 nm is approximately half that produced at 215 nm (see Figure 7.27).

7.5.1 EXPERIMENTAL RESULTS

A typical induced absorption as 215 nm over exposure time (number of pulses) is shown in Figure 7.28 for a high purity-fused-silica (ppb level of impurities) material.

There are two material- and/or process-related parameters that have an effect on the rate at which the absorption develops over time. One is the SiOH content and the

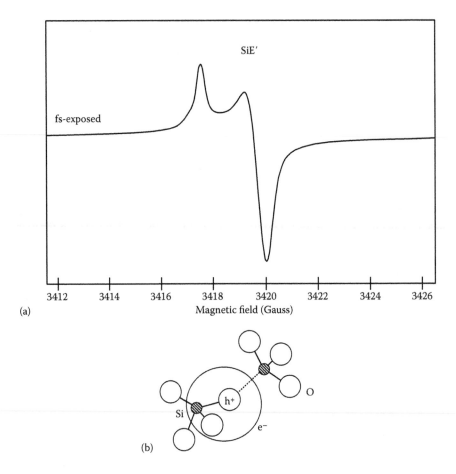

FIGURE 7.27   (a) Characteristic EPR signal of an SiE′ defect that is schematically shown in (b).

other is the entrained molecular H_2 concentration. In the case of high-purity-fused silica, both are related to the process by which it is made, that being a flame hydroly-sis method where the reaction of the silicon-bearing organometallic as it is oxidized to produce silica produces water and hydrogen as the by-products. One can see how the rate of absorption is influenced by these two factors. This absorption is attributed to the SiE′ center, which will be discussed in the next section. The rate appears to follow a [1-exp-(at)] behavior indicative of a first-order reaction, and this will be investigated in the next section.

7.5.2   Mechanisms

The exposure of fused silica to deep UV radiation produces what is termed as a hole trapped on a silicon (SiE′ center) and a hole trapped on a nearby oxygen called a NBOHC. The schematic of these centers is shown in Figure 7.27b. Both of these

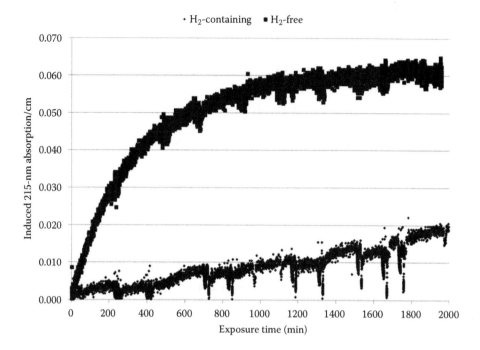

FIGURE 7.28 Example of the induced absorption in high-purity-fused silica produced as a function of number of 193 nm ArF excimer pulses for a number of different OH and H_2 levels.

species have been observed by luminescence in the latter case and both from electron spin resonance. The EPR spectrum is shown in Figure 7.27a.

These centers are thought to form from a self-trapping mechanism within the ideal structure. The self-trapping process requires strong electron-phonon interaction and the magnitude of the energy of the bandgap and additionally on structural disorder. In this case it may be more appropriate to think of the initial formation of a self-trapped exciton that then decays into the E′ and NBOHC; in other words, this can be viewed as resulting in a bond scission. There is an accompanying density change that occurs as a consequence of structural rearrangement as the exciton decays and produces broken bonds. We deal with this in more detail in the photorefractive effect discussion in Chapter 6.

A simple model of the induced suggested by the data contained in Figure 7.28 is the following basically starting from the above mentioned formation of a self-trapped exciton which then decays to the SiE′ from the bond scission. In the absence of any hydrogen, we can write the following equations where P_0 is a measure of the susceptible sites often referred to as strained bonds.[13] These are bonds contained in smaller 3–5 membered Si-O rings.

(a) $h\nu$ X

(b) X → SiE′ + NBOHC

(c) $\quad \dfrac{dE'}{dt} = k_{ex}I(P_0 - E')$

(d) $\quad E' = P_0[1 - \exp(-k_{ex}\,t)]$ \hfill (7.12)

For the case when no significant H_2 concentration is present, this simple form seems to adequately describe curves 1 and 2 in Figure 7.28. However, when H_2 is present two things happen: one is that the rate of absorption is decreased and the other is that the absorption saturation is not apparently ever reached.[16] The explanation put forth for the diminished absorption is that the hydrogen reacts with the NBOHC as pictured in Figure 7.27 to form a -Si- O-SiH that has no absorption in the UV region of interest. If this is true, the absorption would eventually increase with time because the H_2 is being used up, as this behavior has been observed at very long exposures. This means we must add a term to Equation 7.12c to account for reaction with H_2:

$$E' + H_2 = SiH + H.$$

$$\dfrac{dE'}{dt} = k_{ex}I(P_0 - E') - k\left(H_0^0 - E'\right) \hfill (7.13)$$

This would predict that the rate would slow when and if $P_0 > H_0^0$ is achieved. The Araujo paper[16] gives a more elaborate model.

APPENDIX 7A: A PHOTOCONDUCTIVITY STUDY

In Section 7.2.1, the possible role that the mobility of electrons and holes in the respective bands have on the mechanism that produces color centers was mentioned. Relevant to this point is an unpublished study that will be shown here that involved the measurement of the photoconductivity of a number of glasses including silica. Figure 7A.1 shows the schema of the measurement technique used to measure the photoconductivity of the glasses whose composition is shown in Table 7A.1. Some representative results of the measurement for two glasses is shown in Figure 7A.2.

Results are shown for both polarities of the bias voltage, which allowed the conductivity in the valence band and the conduction band to be obtained and compared. It is interesting to note that mobility can differ significantly from one glass to another; that is, the effective mobility is not the same, and further, that there is also unequal mobility in the respective bands. However, the basic intent of the study was to establish that there were continuous extended bands where electrons and hole could travel and to estimate the width of the extended conduction states from the photocurrent measurement as a function of the excitation wavelength. This was done by using a set of absorption cutoff filters, as indicated in Figure 7A.3. The general expression for the dependence on energy is written as

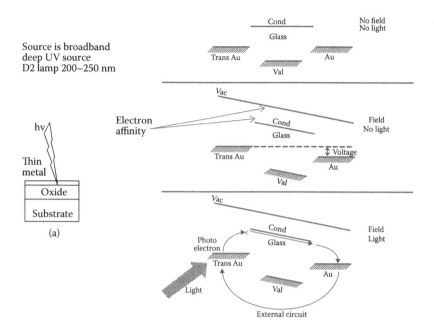

FIGURE 7A.1 Energy band structure schematic of the photoconductive process.

TABLE 7A.1

Glass Compositions of Samples Where Photoconductivity Was Measured

Mol%	JJQ	FD	CZB
SiO_2	70	67.63	53.4
Al_2O_3	22.5	11.01	3.2
B_2O_3	0	9.84	25.1
Li_2O	0	0	3
GeO_2	0	0	15
MgO	0	2.27	0
CaO	0	8.74	0
SrO	0	0.52	0
P_2O_5	7.5	0	0
CeO_2	0	0	0.02
SnO_2	0	0.07	0.07

$$I = A\,(E - \varphi)^2 \tag{7A.1}$$

where E is the excitation energy and φ is the effective work function. If one plots the square root of the photocurrent versus E, the intercept gives the value of φ and the difference between this value and the work function of the metal contact yields the electron affinity (the effective thickness of the extended state conduction band). The electron affinity values for the five glasses measured were in the range of 0.1–0.25 eV.

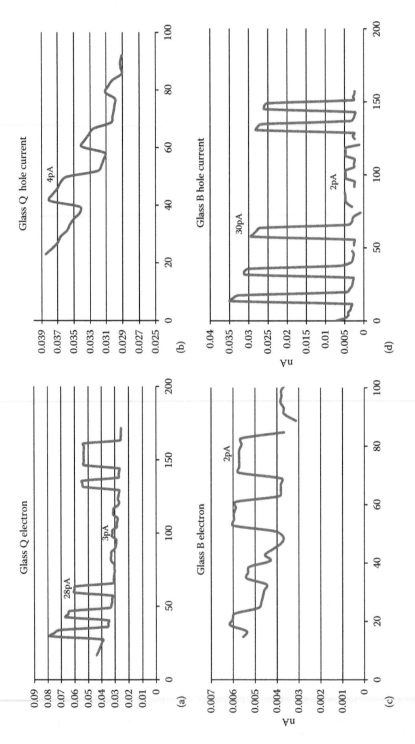

FIGURE 7A.2   Experimental results for two glasses (a and b) CZB and (c and d) FD in Table 7A.1.

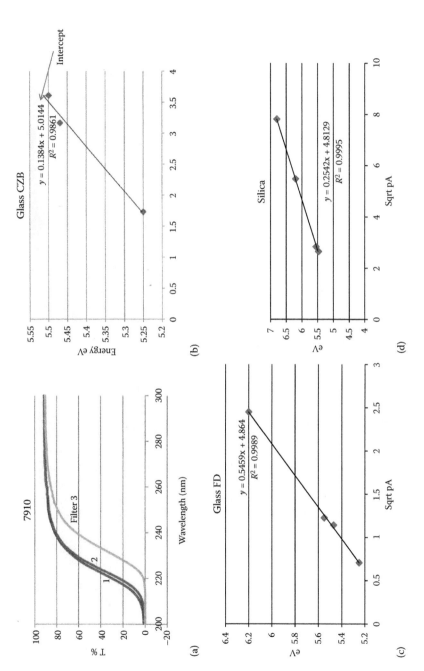

FIGURE 7A.3 (a) Cutoff filters used to determine the wavelength dependence of the photocurrent, (b) the excitation energy versus the square root of the photocurrent to determine the estimate of the electron affinity of glass CZB, (c) the same for glass FD, and (d) the same for silica.

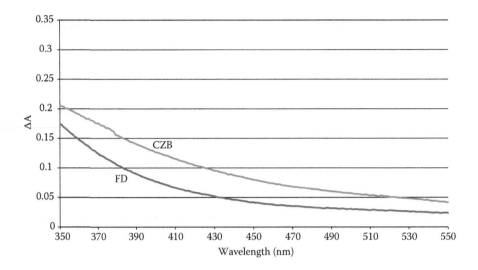

FIGURE 7A.4   UV-induced absorption (ΔA) of glasses CZB and FD.

Relevant to the solarization phenomenon we also subsequently tried to see if there was any correlation between the mobility of the electrons and/or holes with the extent of induced absorption, as mentioned in the last section. The direct correlation or lack thereof is made difficult because of trying to establish similar exposure conditions. We chose two glasses (CZB and FD in Table 7A.1) that had approximately the same initial absorption coefficient at the 248-nm laser wavelength. The photocurrent results are shown in Figure 7A.2. The induced absorption curves for the two glasses are shown in Figure 7A.4. Clearly, glass CZB exhibits a larger degree of induced absorption and has much lower electron mobility but much higher hole mobility; glass FD has the opposite behavior. The conductivity is proportional to the product of the carrier density and the mobility. If the carrier concentration can be assumed equal because of the equal absorption coefficient at the excitation wavelength, then we are essentially comparing mobility or mean free path. The correlation, if any, is ambiguous and depends on how one imagines the process to proceed. If the electron trapping is dominant then a higher mobility would imply less trapping and thus less induced absorption, which appears to be the case. Clearly, this conclusion is based on one experiment and is not enough to draw a definite conclusion. On the other hand, it does seem to suggest that the mobility of the photo-produced electron or hole might be important.

REFERENCES

1. C. Kittel, *Introduction to Solid State Physics*, Third Edition, John Wiley & Sons, New York, 1966.
2. A.J. Dekker, *Solid State Physics*, Prentice Hall, Englewood Cliffs, NJ, 1959.
3. D.L. Griscom, *J. Non-Cryst. Sol.* 6, 275, 1971.
4. G.H. Sigel, *J. Phys. Chem. Sol.* 32, 2373, 1971.

5. H. Yun and P.J. Bray, *J. Non-Cryst. Solids* 27, 363, 1978.
6. N.F. Mott, Electrons on glass, *Nobel Lectures*, December 8, 1977.
7. L.B. Glebov, Laser induced damage in optical glasses, *SPIE* 4347, 2001.
8. T. Bates, Ligand field theory, Chapter 5, in *Modern Aspects of the Vitreous State*, Vol. 2, J.D. Mackenzie (ed.), Butterworth, Washington, DC, 1962.
9. D. Ehrt, *C. R. Chim.* 5, 679, 2002.
10. H. Hosono and Y. Abe, *J. Non-Cryst. Sol.* 125, 98, 1990.
11. M.J. Dejneka et al., U.S. Patent Application.
12. D.C. Allan et al., *Opt. Lett.* 21(24), 1960, 1996.
13. N.F. Borrelli et al., *J. Opt. Soc. B.* 14(7), 1997.
14. N.F. Borrelli, D.C. Allan, and R.A. Modavis, *J. Opt. Soc. Am.* 16(10), 1672, 1999.
15. H. Nishikawa et al., *J. Non-Cryst. Sol.* 179, 1994.
16. R.J. Araujo et al., *Proc. SPIE* 3424, July 1998.

8 Photochemistry in Porous Glass

What a tangled web we weave ...

Sir Walter Scott

8.1 INTRODUCTION

8.1.1 POROUS GLASS

This chapter will discuss a much different approach to photosensitive processes in glass, or perhaps the more appropriate word would be "on" glass, since it will be more of a surface reaction. The glass we will describe is a unique one that has been around for over 80 years.[1] What makes it a somewhat interesting story is that it is a glass produced in an intermediate stage to the intended product; namely, a high 97% silica glass melted in a conventional glass-making manner. High silica glass, like fused silica, is formed by a vapor deposition process and not by conventional melting because the melting temperature is > 1800°C. The unique invention[1,2] was to melt an alkali borosilicate glass and then take advantage of a known effect in certain glass composition systems, which is the thermodynamic phenomenon of phase separation. This is the situation where the glass is initially formed as a homogeneous glass because of the relatively rapid cooling from the melt. In other words, the true equilibrium structure of the glass is not realized. However, upon a subsequent heat treatment ~500°C, the glass is allowed to approach a situation where it physically separates into two phases, one silica-rich and the other borate-rich. The composition region of the ternary $Na_2O/B_2O_3/SiO_2$, which circumscribes this behavior, is shown in Figure 8.1 and a simple phase diagram is shown in Figure 8.2.

Although the phase diagram tells you what composition split would occur for a given temperature, it does not say anything about the morphology of the separated phases. By morphology, we mean the shape length and size of the separated phase. To understand this aspect, it is important to refer to the thermodynamic free energy diagram at a given temperature, as shown in Figure 8.2. Spinodal decomposition[3] refers to when the two phases are completely interpenetrating as Vycor™ is, which will be discussed in Section 8.2.1.

The spinodal phase separation is driven as a consequence of $d^2g/dc^2 < 0$, but the specific scale of the physical morphology of the separation depends on a number of other factors. A significant and relevant aspect of the porous Vycor glass (PVG) is that it phase separates in a spinodal fashion (totally interpenetrating borate phase in the

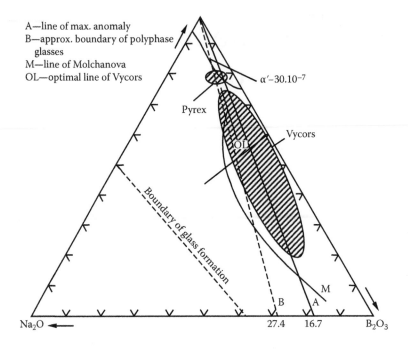

FIGURE 8.1   Ternary $Na_2O/B_2O_3/SiO_2$ phase diagram indicating the Vycor region of composition space.

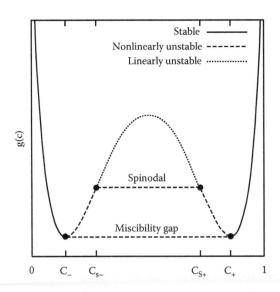

FIGURE 8.2   Idealized free energy versus concentration graph indicating the different regions defining the spinodal and miscibility regions defined by C_s and C_i, respectively.

silica-rich matrix) at a point where the average width of the serpentine channel is <1 nm. The framework of the approach developed by Cahn and Hilliard[4] is to incorporate the concentration fluctuation into a term in the free energy. This allows us to use the classical diffusion equation with additional terms reflecting the spatial concentration fluctuation and the free-energy dependence on the concentration that allows counterdiffusion. The next step in the process is relevant to the photosensitive properties that will be described in subsequent sections. The borate phase is very soluble in water or weak acid so the complete interpenetrating phase can be dissolved away, leaving the porous silica-rich structure intact. This leaching process is done in a temperature-controlled slightly agitated bath in order to obtain a uniform pore size structure throughout the body. The drying process is also carefully controlled to prevent cracking.

It was after this step of creating the high silica porous body that the originally intended product was to be made by heating the porous glass body to ~1200°C where the nanometer-size pores collapse, leaving a dense 97% silica glass with physical and thermal properties resembling fused silica.[2] Our interest in this chapter will be in the utilization of the intermediate porous state as a host for photosensitive activity. Figure 8.3 shows an SEM of the resulting porous structure after leaching and Figure 8.4 shows the measured pore size distribution.

The following list gives other pertinent physical and thermal properties of the resulting structure, the most important of which is the total internal surface area measured as 200 m^2/g or 130 m^2/cm^3. An equally important relevant property is that the surface area is chemically active through the presence of a high concentration of Si-OH-terminated bonds.

Important Properties of PVG

- Density ~1.5 g/cc
- About 28% pore volume
- Surface area ~200 m^2/g
- Average pore diameter 40–50 angstroms
- Composition
    - –96% SiO_2
    - –~3% B_2O_3
    - –~0.7% Al_2O_3, ZrO_2, etc.

Less Than 400 ppm Na

- Heat treatment above 1000°C closes pores
- ~13% linear shrinkage
- Silica-like glass properties
- Same 96% SiO_2, 3% B_2O_3 composition
- CTE ~7.5 × 10^{-7}/K
- Density, 2.18 g/cc
- Refractive index, 1.458
- Annealing point, 1020°C
- Decreased tendency to devitrify compared to fused silica

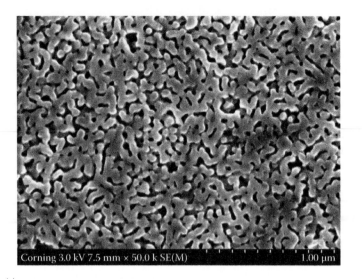

(a)

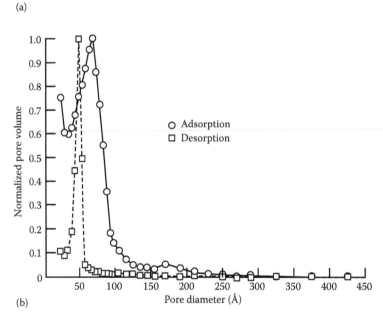

(b)

FIGURE 8.3 (a) SEM of the leached Vycor and (b) measured pore volume versus pore size distribution.

As a result of these bonds, a variety of metal oxides and organometallic compounds can be impregnated by a solution method and be held relatively stable at room temperature through either chemisorbed or physisorbed mechanisms. It should be mentioned that the porous Vycor has been used as a catalyst support in the past, taking advantage of the highly reactive surface. Some important organo-metallic compounds that were utilized are shown in Figure 8.4.

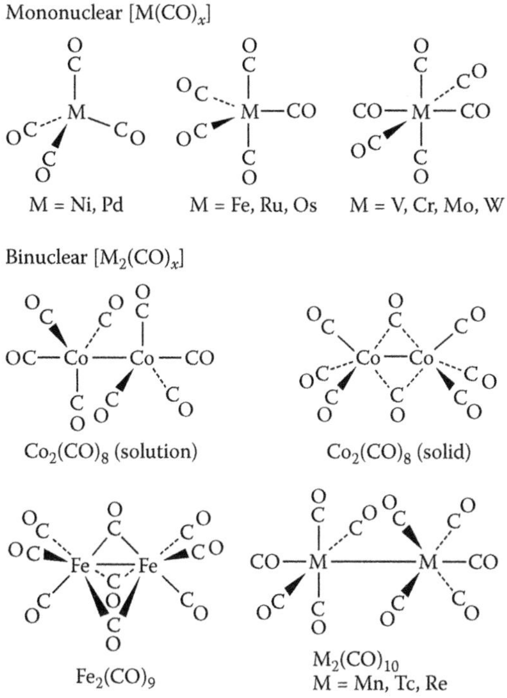

Mononuclear [M(CO)$_x$]

M = Ni, Pd   M = Fe, Ru, Os   M = V, Cr, Mo, W

Binuclear [M$_2$(CO)$_x$]

Co$_2$(CO)$_8$ (solution)   Co$_2$(CO)$_8$ (solid)

Fe$_2$(CO)$_9$

M$_2$(CO)$_{10}$
M = Mn, Tc, Re

FIGURE 8.4 Examples of organometallic molecules that can be solution-loaded into the porous network of PVG.

8.2 THE PHOTOSENSITIVE PROCESS

8.2.1 PHOTOSENSITIVE SAMPLE PREPARATION

The following process is unique in that one can incorporate the oxides of almost any of the transition metal ions into silica. If we were to follow the conventional process of glass melting it would be to initially add the transition metal ion into the glass as a batch material before melting. But even with this approach there is no guarantee that the oxide of the metal ion would appear as a separate phase within the glass. The photosensitive processes as will be described here are based on the utilization of the chemically active high-surface area's porous Vycor glass as the host for photosensitive molecules such as the transition metal ion carbonyls. Because of the high concentration of Si-OH groups on the surface of the porous narrow channel network, the photosensitive molecules remain stable at room temperature in the absence of light. The stability is reasoned to be the result of either physisorbed or chemisorbed reaction with the OH groups on the inner walls of the porous structure. The initial transmission of the porous Vycor glass (PVG) is shown in Figure 8.5. One can see the excellent transmission of the deep UV, which, as we will see, plays an important role in photocatalytic activity.

Prior to loading, the porous sample was heated in air to 600°C to remove all the carbon impurities that are trapped within the glass over time by exposure to

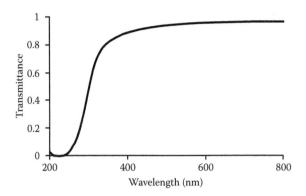

FIGURE 8.5 Spectral UV-vis transmission of PVG.

the ambient. In most cases the photosensitive molecule is in solid form and then is dissolved in a suitable solvent, the concentration of which is usually dictated by the maximum solubility. The amount that is actually loaded into the sample is not dependent on the solution concentration beyond a certain point but rather by the maximum amount that could be absorbed; that is, the number of active sites on the porous channels. The porous sample was then allowed to soak in the solution for a sufficient time to penetrate the entire thickness of the sample, which for a 1-mm thick sample took about 1 h. In some cases for particular molecules the loading was done by sublimation; that is, the sample was placed over the powder and gently heated. The loaded sample was then allowed to dry in air or at 100°C depending on the stability of the particular photosensitive molecule.

8.2.2 PHOTOSENSITIVE MOLECULES

Figure 8.4 lists a partial but representative number of the candidate photosensitive molecules and structures that were initially studied that were loaded into the PVG and then exposed to light in the spectral region listed. These are almost all metal carbonyls, which are a well-known set of photosensitive molecules.[5,6] There are other molecules in the M-(cylcopentadiene)$_2$Cl$_2$ family which were used that also are known to be photosensitive. By photosensitive molecule, we mean that exposure to light of the appropriate wavelength can strip ligands from the metal, leaving it in a reactive state. The photosensitive nature of the organic-metallic compound studied here appears not to measurably change when the molecule is entrained in the porous structure. In general one could approximate the situation as the molecule being in a low temperature condition, which could contribute to greater stability.

8.2.3 EXPOSURE SOURCES AND METHODS

The exposure source was usually a visible wavelength laser, He-Ne 633 nm and 560 nm of an Argon-ion laser 488–512 nm or a filtered Hg-Xe lamp. The wavelength was selected to match the absorption spectrum of the particular photosensitive molecule being used. In some cases, as we will see, a focused beam was used to write patterns whereas in other situations a photomask was used with a flood exposure. As we discuss each application, the intensity level will be specified. In all cases the exposure was done in air at room temperature since the photosensitive molecules used were not particularly air-sensitive.

8.3 PHOTOSENSITIVE APPLICATIONS

Six examples will be given of how the impregnation of a photosensitive agent into the porous glass structure leads to a variety of optical effects produced by exposure to light. Because of the nanometer scale of the pores and further to the subsequent consolidation one obtains dense glasses that extends the photosensitive range and property versatility of the conventionally melted glasses covered in Chapter 2. Another way of expressing this definition is that at the end one could not distinguish how the optically induced effect was produced.

The examples of the photosensitive effects to be described are all a result of the impregnation of a photosensitive molecule into the porous structure followed by an exposure to light to produce a permanent change in a physical property. In the first case it will be the production of a patterned magnetic phase. In the next two examples it will be the production patterned induced refractive index changes. The following two will be methods describing a way to photopattern quantum dots (CdSe and GaAs). The last example is quite different in showing how one can produce CH_4 from CO_2 using a photocatalyst using PVG as the host substrate.

8.3.1 THE PHOTO-INDUCED MAGNETIC PHASE

8.3.1.1 Sample Preparation

In this example,[5,6] photochemical reactions were observed in a porous glass impregnated with iron pentacarbonyl ($Fe(CO)_5$). The porous glass samples, typically $25 \times 25 \times 2$ mm^3 were loaded with iron pentacarbonyl either by dipping the sample in a solution (usually dilute CH_2Cl_2) of the $Fe(CO)_5$ for a few minutes or by exposing the sample to the $Fe(CO)_5$ vapor. The glass was typically exposed to UV light (100W Hg arc 1–10 minutes) and then subsequently heated in air, resulting in a photoreaction that produced a magnetic iron oxide species either magnetite (Fe_2O_3) or maghemite (γO_3). Visible wavelength could also be used since the absorption of the $Fe(CO)_5$ extended using a number (e.g., krypton and Ar-ion lasers). The photoreaction was attributed to the breaking of one or more of the ironcarbonyl bonds, leaving a reactive iron species which, in turn, reacts with the silanol (Si-OH) groups on the surface of the pores. Subsequent heating oxidizes the remaining carbonyl groups, leaving a form of iron oxide which is bound to the surface of the glass (see Figure 8.6). The structure was then consolidated to a dense glass by heating to 1200°C. The latter temperature was sufficient to consolidate the sample by closing the pores because the glass was soft enough to flow. Figure 8.7 shows a TEM image of the iron oxide particles in the consolidated silica matrix.

FIGURE 8.6   Schematic representation of the photoreaction of $Fe(CO)_5$ inside the PVF channel.

FIGURE 8.7   TEM image of magnetic particles formed in the PVG after consolidation. (From N.F. Borrelli, D.L. Morse, and J.W.H. Schreurs, *J. Appl. Phys.* 54(6), 3344, 1983.[7])

8.3.1.2   Magnetic Properties

In the following, a brief summary of the results is given for the magnetic properties of the iron oxide impregnated and consolidated Vycor samples.[7] Figure 8.8 shows the RT hysteresis loops for two samples that were UV-exposed and then consolidated at 1200°C. The measured magnetic parameters; namely, the saturation magnetization (emu/g) and the coercive field (Oe) that measures the field to reverse the magnetization, are summarized in Table 8.1.

Note that the measured coercive field is unusually large, especially for magnetite Fe_3O_4 and/or maghemite γFe_2O_3.[8] Later in this section we will see the possible reasons for this occurring in very small particles will be discussed. Another meaningful

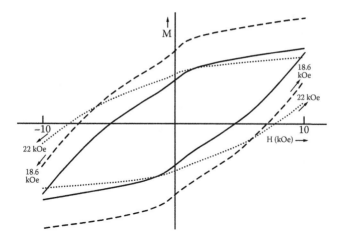

FIGURE 8.8 Hysteresis measurements for two photolyzed $Fe(CO)_5$-loaded PVF samples and then consolidated at 200°C. (From N.F. Borrelli, D.L. Morse, and J.W.H. Schreurs, *J. Appl. Phys.* 54(6), 3344, 1983.[7])

TABLE 8.1
Summary of the Magnetic Parameters
for the Sample Measured in Figure 8.8

Sample	H_{max} (kOe)	M_{10k} (emu/g)	H_c (Oe)
1	10	0.089	5940
	18.6	0.124	8484
2	22	0.046	9090

Source: N.F. Borrelli, D.L. Morse, and J.W.H. Schreurs, *J. Appl. Phys.* 54(6), 3344, 1983.

---

experimental result was the strong temperature dependence of the coercive field, which is shown in Figure 8.9.

The development of the ferrimagnetic phase within the pores was studied by measuring the magnetic hysteresis loops as a function of temperature after exposure at temperatures below the consolidation temperature. Here, one can study essentially the growth of the amount of the magnetic phase and the particle size. The estimate of the amount of magnetic phase being produced in a glass can be obtained by assuming the phase is the γFe_2O_3 form with a saturation magnetization of 80 emu/g.[7] Using the average values obtained experimentally leads to a value ~1 mg magnetic phase/gram of glass.

In the foregoing discussion it was shown that one can produce a ferrimagnetic phase in porous glass upon consolidation by exposing the sample to light after impregnation with iron carbonyl. The magnetic properties obtained are unusual in two ways: first, the very high coercive force that can be obtained, and second, the very strong

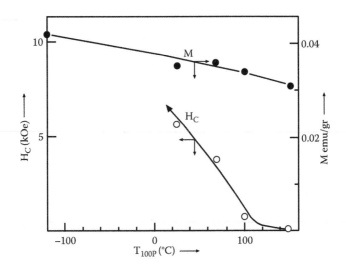

FIGURE 8.9 Dependence of magnetization and coercive field as a function of temperature of exposed and consolidated Fe(CO)$_5$-loaded PVG. (From N.F. Borrelli, D.L. Morse, and J.W.H. Schreurs, *J. Appl. Phys.* 54(6), 3344, 1983.[7])

temperature dependence of the coercive force. It is known that the magnetic behavior of very small particles are size-dependent.[8,9] This is due to the volume-dependent factors that are involved, primarily magnetocrystalline anisotropy energy (forces internal to hold the electron spins in a given crystalline direction); K measures the magnitude of this effect, demagnetization energy (energy contained in the field), and strain anisotropy energy (effect of external pressure).

When the size of the particle is small enough the crystalline anisotropy energy, KV, becomes comparable to the thermal energy, kT. Here, the value of the anisotropy constant is in ergs/cm^3 and V, the volume, in cm^3. When this is the case, the thermal vibrations randomize the magnetization directions of the particles so that in the absence of an applied magnetic field the overall magnetization is zero. However, because the magnetic moment of an individual particle is still quite large compared to that of a single iron ion, the coupling to an applied magnetic field is strong, so that moderately strong fields can cause a great deal of spin alignment. The magnetization versus applied field curve for such superparamagnetic particles[8] are shown in Figure 8.10.

There is no remnant magnetization, but saturation already occurs for moderately strong fields. An estimate of the critical size for the onset of superparamagnetic behavior can be made from the expression $KV/8 = 2.2kT$ where K is the magnetocrystalline anisotropy constant (energy/cm^3), V is the volume, k is Boltzmann's constant, and T is the absolute temperature.

Electron microscopy has shown that for a consolidated Fe(CO)$_5$-impregnated sample, which was photolyzed by UV, the distribution of particles is relatively uniform over the radius range of 5–10 nm. It appears that the observed rapid decrease in the coercive force as the temperature is increased is actually due to the temperature dependence of the critical size. The observed coercive force value in excess of 10 kOe is quite unusual, except in

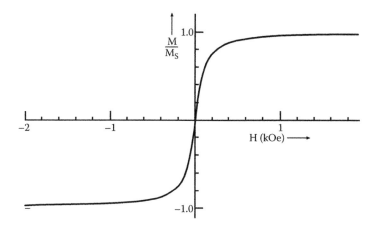

FIGURE 8.10   Example of the magnetic behavior in the superparamagnetic state $K \sim kT$. (From N.F. Borrelli, D.L. Morse and J.W.H. Schreurs, *J. Appl. Phys.* 54(6), 3344, 1983.[7])

the case of rare-earth cobalt metals. Even for these materials, 10 kOe is the typical value. As mentioned above, there are three sources of anisotropy that control the coercive force in single-domain ferrimagnetic particles. The high coercive force is proposed to result from a large magnetocrystalline anisotropy of the very small particles produced in the manner. Somehow the way the magnetic particles form in the pores in the PVG structure and then consolidated must produce a material with an unusually high magnetocrystalline anisotropy, perhaps combined with a pressure induced effect. At this point, it is conjectural as to the source of the very large coercive field.

8.3.2   INDUCED REFRACTIVE INDEX CHANGES

The list of important properties of PVG (given in Section 8.1.1) provides a partial listing of photosensitive organometallic compounds. In many of these cases, when these molecules are loaded into porous Vycor and exposed to light and heated, a permanent refractive index change can be induced.[5,6,10–12] Because the photosensitive reaction is occurring in nanometer-size pores, the obtainable resolution of index change is limited by the resolution of the exposure beam. Some examples of the types of optical effects that will be described will be shown. In the first example, the organometallic molecule loaded into the glass was $Mn_2(CO)_6$, which can be photolyzed by red light. Figure 8.11 shows an index pattern induced as a result of 633-nm laser exposure through an exposure mask and the corresponding refractive index pattern as measured by a ZYGO interferometer. Figure 8.12 shows a diffraction pattern produced by a diffraction grating made by exposing the loaded glass through an exposure mask.

The photoreaction of the $Mn_2(CO)_{10}$ is similar to that described in the previous section involving $Fe_2(CO)_5$ where the light (633 nm this case) strips a CO, leaving an unsaturated ligand that in turn attaches to the pore wall through the SiOH silanol group. Subsequently, after heating in air and then at 1200°C consolidation, we are left with Mn-oxide, which raises the refractive index in the exposed region of the

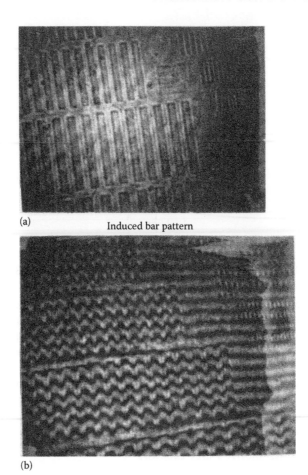

(a)   Induced bar pattern

(b)

FIGURE 8.11 Two examples (a and b) of refractive index patterns produced by visible light exposure of $Mn_2(CO)_6$-loaded PVG and then consolidated. (From N.F. Borrelli and D.L. Morse, *Appl. Phys. Lett.* 43(11), 992, 1983.[5])

glass. The simple formulation would follow a linear relationship where V_f, the volume fraction of the metal oxide,

$$n_{eff} = n_{Mn\text{-}oxide} V_f + (1 - V_f) n_{SiO2}$$

$$\Delta n = V_f(n_{Mn\text{-}oxide} - n_{SiO2}) \tag{8.1}$$

From the interference pattern measured by a ZYGO interferometer we can estimate the induced index change to be in the order of 0.003, which would allow an estimate of a volume fraction of metal oxide of ~0.4% (vol).

We can also use the induced refractive index mechanism to make a gradient index lens.[13] The exposure through a small round opening in an otherwise opaque mask produces a quasi-parabolic refractive index profile that satisfies the imaging

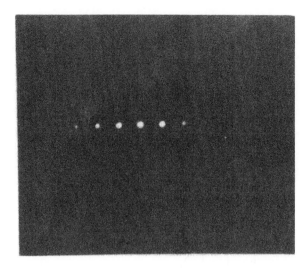

FIGURE 8.12   Diffraction pattern obtained from phase grating written in $Mn_2(CO)_6$-loaded PVG and then consolidated. (From N.F. Borrelli and D.L. Morse, *Appl. Phys. Lett.* 43(11), 992, 1983.[5])

condition of the refractive rays. A microscope photo of the exposed lens area and the image formed by the lens is shown in Figure 8.13.

A related study[10] used as the photosensitive agent cyclopentadienal titanium dichloride, abbreviated as titanocene dichloride, which was solution-loaded into the PVG. The proposed photomechanism is shown in Figure 8.14 and is similar to that shown above for the carbonyls.

The exposure wavelength range for the photoexcitation is shown in Figure 8.15. This method was used to provide a way to write optical structures such as optical waveguides, optical splitters, and diffraction gratings. The examples of the structures

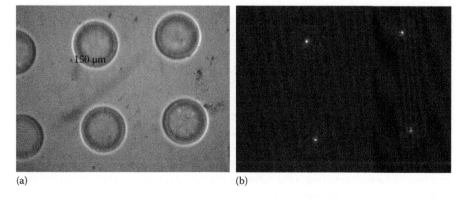

(a)   (b)

FIGURE 8.13   A gradient index lens made by exposure of $Mn_2(CO)_5$-loaded PVG by exposure through a pinhole produces a quasi-parabolic index profile: (a) photo of the exposed regions after heating and (b) image formed by lenses at the focal plane.

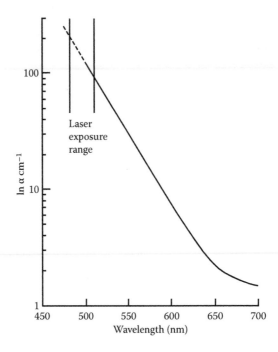

FIGURE 8.14 Schematic of the reaction process using cyclopentadienal-titanium-dichloride-loaded PVD. (From N.F. Borrelli, M.D. Cotter, and J.C. Luong, *IEEE J. Quantum Electron.* 22(6), 1986.[10])

FIGURE 8.15 Transmission spectrum of a titanocene-loaded PVG sample indicating the wavelength exposure region. (From N.F. Borrelli, M.D. Cotter, and J.C. Luong, *IEEE J. Quantum Electron.* 22(6), 1986.[10])

to be shown are all in the consolidated state. The optical setup for the writing is shown in Figure 8.16. The writing laser used to create the waveguide structures was a 400-mW argon-ion laser focused with an effective 0.1-NA objective. An example of a 1-4 waveguide splitter is shown in Figure 8.17. The guide dimensions were estimated to be on the order of 50 μm wide and 80 μm deep. More elaborate waveguide structures are shown in Figure 8.18. On the left in Figure 8.18, is the design of the structure and adjacent to that is the actual guide that was made, and the photos on the right show the light in the guide coupling from the straight guide (top) to the circular guide (bottom).

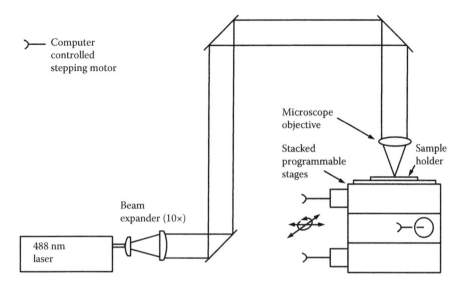

FIGURE 8.16   Optical arrangement for writing waveguide structures in titanocene-loaded PVG.

8.3.3   OTHER PHOTOSENSITIVE/VYCOR PHENOMENA

8.3.3.1   The Preparation of the Semiconductor Nanophase

There is another phenomenon that uses light and the porous Vycor structure to produce semiconductor nanophase.[14] Optical effects exhibited by semiconductor microstructures and microcrystals are of considerable interest both for the underlying physics and the potential application in the area of nonlinear optical devices. Methods of fabricating these microstructures vary depending on the material and the nature of the desired effect. Here we describe a method of preparing three dimensionally quantum confined structures of II-VI and IV-VI compounds utilizing a porous glass structure. The method involves a gas phase reaction of appropriate compounds impregnated in a porous glass support. This potentially has the advantages of easier preparation, greater stability, and the possibility of photopatterning through photoinitiating the reaction. In addition, the use of a porous glass host affords the possibility of multiple optical structures to be fabricated in one piece of glass.

For the II-VI compounds such as CdS and CdSe, the method was a gas phase reaction of the chalcogenides with the suitable metal ions impregnated in the porous glass. This was accomplished by exposing the porous glass (prefired at 650°C) to an aqueous solution containing variable concentrations of metal salts. Metal chlorides were used for Zn and Cd and metal (II) acetates were used for the IV-VI compounds such as PbS and PbSe. The source of the chalcogenide was derived either directly from the gaseous hydrogen chalcogenide or in situ from the thermal decomposition of thiourea (selenourea) coimpregnated in the porous glass. For the photoinitiated reaction the starting material was N,N-dimethyl selenourea.

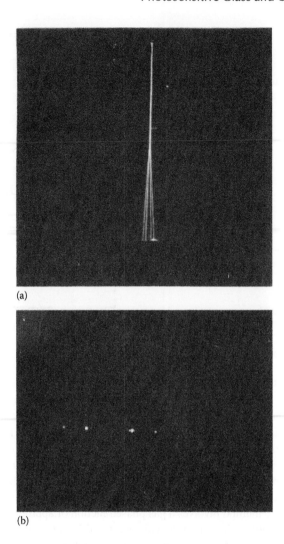

(a)

(b)

FIGURE 8.17 An example of a waveguide 1-4 splitter structure using the arrangement described in Figure 8.16. (a) Top view of a splitter and (b) end view of four output beams. (From N.F. Borrelli, M.D. Cotter, and J.C. Luong, *IEEE J. Quantum Electron.* 22(6), 1986.[10])

The pore size does provide an upper limit to the crystallite dimension, but evidence for narrow size distribution was not seen. Direct evidence for the crystalline structure produced within the porous glass was obtained from x-ray diffraction for PbS and MoS2.

As an example, the results of the optical absorption spectra of CdSe are shown in Figure 8.19. In this case the crystallite size is sufficiently small to exhibit a 0.3-eV absorption edge shift relative to the bulk. Comparison to CdSe thermally developed in glass shows a similar edge to that treated at 575°C. Although electron

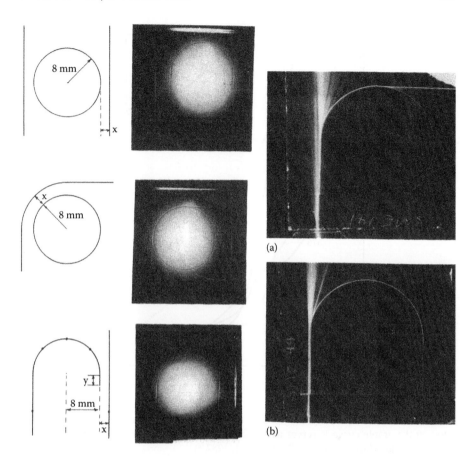

FIGURE 8.18 Examples of more elaborate functional waveguide structures. The left-most figures show the intended tracks, the images in the center show the actual written patterns propagating light, and the right-most column shows an expanded view of the light propagation. (From N.F. Borrelli, M.D. Cotter, and J.C. Luong, *IEEE J. Quantum Electron.* 22(6), 1986.[10])

microscopic measurements were not made on either of these two samples, an estimate of the mean crystallite size can be made based on the edge shift relative to the 600°C treatment where the size was measured as 3 nm. Using the theoretical result that the bandgap shift is proportional to $1/a^2$, with "a" being the crystallite diameter, a value of about 2 nm is obtained. The wavelength corresponding to the maximum luminescence is 400 nm, suggesting that the average diameter may even be smaller. The absence of distinct banding in the absorption spectrum is likely due to the broad crystallite size distribution. The optical absorption spectrum for the CdSe formed at 0.1 M Cd salt concentration; see curve B in Figure 8.19.

A connection to the photoeffects discussed above is the result that one can pattern CdSe by exposing the loaded porous glass (Cd-salt and urea) to a CO_2 laser through

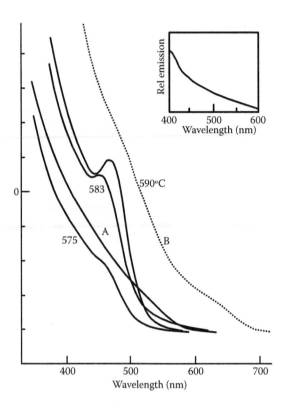

FIGURE 8.19 Absorption spectra of CdSe-loaded PVG glass treated at the temperature noted on the graph; curve B serves as a reference to the band edge of bulk CdSe. (From J.C. Luong et al. *MRS Symposium Proc.* 75, 671, 1987.[14])

a patterned photomask. The 10.6-μm laser radiation is absorbed by the glass in the openings in the mask, thus heating that local region which thermally drives the reaction. An example of the patterning is shown in Figure 8.20.

8.3.3.2 Photocatalysis

An unlikely application of the porous Vycor glass is the utilization of the porous structure as a photocatalytic support for the conversion of CO_2 to CH_4.[15,16] The research focuses on the discoveries made during photocatalytic experiments using porous Vycor as the support for a tungsten hexacarbonyl precatalyst, $W(CO)_6$. The reaction was done under mixed visible/UV irradiation with the eventual goal of converting catalytic performance into the visible spectrum. Catalyst-loading was performed through wet impregnation of the slides with a hexane solution of $W(CO)_6$. The bare PVG sample was transferred into a cell, shown in Figure 8.21, and dried under dynamic vacuum for 30 minutes. Exposures were made with an Hg lamp 6281. Lamp average intensities were between 20 and 35 mW and between 120 and 145 mW, respectively.

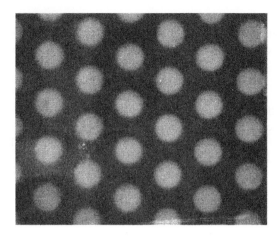

FIGURE 8.20 CO_2 laser patterning of the regions of CdSe-loaded PVG by local heating through a mask. (From J.C. Luong et al. *MRS Symposium Proc.* 75, 671, 1987.[14])

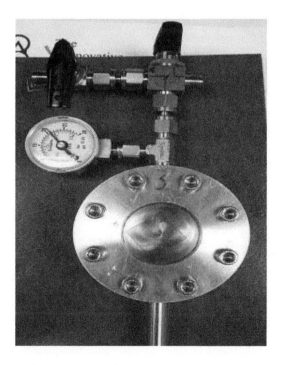

FIGURE 8.21 Experimental cell used to study the photocatalytic reactions in catalyst-loaded porous Vycor glass, as described in the text.

The simple reaction can be written formally in the following way, although it is not free-energy-justified and all the intermediate steps are suppressed. Of course, what a catalyst does is somehow get around the equilibrium thermodynamics.

$$CO_2 + H_2O + hv + catalyst \rightarrow CH_4 + H_2 + O_2 \tag{8.2}$$

Needless to say, the real reaction is quite complex and appears to occur in stages, the first being the UV-driven stripping of a CO ligand from the $W(CO)_6$. CO is indeed one of the products measured, as shown in Figure 8.22a. The second phase of the reaction is the production of hydrogen, as indicated from the results shown in

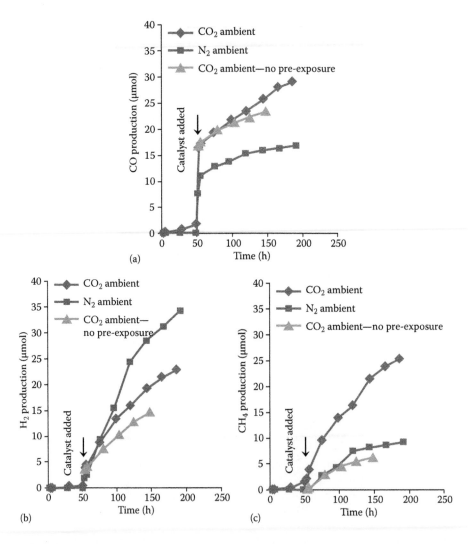

FIGURE 8.22 MS/GC measured amounts of listed gas as a function of UV exposure time: (a) CO, (b) H_2, and (c) CH_4.

TABLE 8.2
Typical Amounts of the Listed Gases after >200 h of UV Exposure

GC Analysis	Catalyst Concentration (mmol)			
Ambient Gas (ppm)	1	10	20	40
H_2	860	3300	4300	6100
O_2	ND	ND	ND	ND
N_2	530	840	150	320
CO	1900	14,000	20,100	25,000
H_2O	990	1000	1300	1900
CH_4	1400	1300	380	80
Ethane	260	320	110	ND
Propane	80	110	40	10
Butane	ND	ND	ND	ND

Figure 8.22b. This hydrogen is assumed to come from the reaction of the unsaturated tungsten species with the SiOH groups at the surface of Vycor and/or adsorbed water. No O_2 was detected, as suggested by the equation on the previous page. Nonetheless, the ultimate stage did produce CH_4, as shown in Figure 8.22c. A typical MS/GCP output is shown in Table 8.2. This is an example of how potentially active the surface of the Vycor is, both chemically and photochemically.

REFERENCES

1. H.P. Hood, U.S. Patent No. 2,106,744, 1934.
2. T.H. Elmer, *J. Am. Ceram. Soc.* 62, 513, 1983.
3. D.A. Porter and K.E. Easterling, *Phase Transformations in Metals and Alloys*, Second Edition, Taylor & Francis, New York, 1981.
4. J.W. Cahn and J.E. Hilliard, *J. Chem. Phys.* 28, 258, 1958.
5. N.F. Borrelli and D.L. Morse, *Appl. Phys. Lett.* 43(11), 992, 1983.
6. H.D. Gafney and D. Sunil, *J. Am. Chem. Soc.* 131, 14768, 2009.
7. N.F. Borrelli, D.L. Morse, and J.W.H. Schreurs, *J. Appl. Phys.* 54(6), 3344, 1983.
8. J. Smit and H.P.J Wijn, Philips Technical Library, Eindhoven, 165, 1965.
9. S. Chikazumi, *Physics of Magnetism*, John Wiley & Sons, New York, 1966.
10. N.F. Borrelli, M.D. Cotter, and J.C. Luong, *IEEE J. Quantum Electron.* 22(6), 1986.
11. M. Anpo and T. Matsuua (eds.), *Photochemistry on Solid Surfaces*, Elsevier, New York, 1989.
12. N.F. Borrelli and J.C. Luong, *Proc. SPIE* 806, 104, 1987.
13. N.F. Borrelli, *Microoptics Technology*, Second Edition, Chapter 3, Marcel Dekker, New York, 2005.
14. J.C. Luong et al. *MRS Symposium Proc.* 75, 671, 1987.
15. H.D. Gafney, *Photodeposition in Glasses*, PN, 1998.
16. H.D. Gafney and E.C. Look, *J. Phys. Chem. A* 117(47), 12268, 2013.

TABLE 8-2

9 Polarizing Glass

I can see clearly now ...

Song lyric, Johnny Nash

9.1 INTRODUCTION

Although polarizing glass is not a photosensitive glass as we defined it initially in this book, nonetheless it is a very close relative in that it involves the manipulation of Ag nanoparticles to produce a polarizing effect. In one important example to be covered below; namely, the infrared polarizer Polarcor™, the polarizing property is derived directly from a photochromic glass composition.[1] The major application for this product is for optical isolators for lasers in a communication network. The optical isolator (polarizer + magneto-optic crystal) prevents any light feedback into the laser cavity to the level of >50-db rejection.[2] In the other example it involves glass compositions where larger Ag nanoparticles can be produced, thus, making it easier to elongate the particle to produce the required anisotropy in absorption. In another sense the inclusion of this topic is apropos because the application of polarizing devices is ubiquitous (e.g., polarizing sunglasses).

9.1.1 HISTORY AND BACKGROUND

The production of light polarizing materials in the form of plastic sheet is a well-known art. The production of glass polarizers is less well known. However, because plastic materials suffer for many applications because of their inherently low surface hardness, relatively high moisture susceptibility, and limited chemical and thermal durability—all areas in which glass performance excels—glass is the material of choice for many optical applications and much effort has gone into developing all-glass polarizers.

Edwin Land,[3] the inventor of plastic sheet polarizers and the first to commercialize them (Polaroid brand products), also experimented with glass polarizers made by elongating (stretching) metal particles suspended in a glass matrix. He reported unusual color effects as well as light polarization, but was apparently unsuccessful at producing a commercial glass-based polarizer product.

Stookey and Araujo,[4] experimenting at Corning Incorporated (then Corning Glass Works), produced strips of polarizing glass based on elongated nanoparticles of silver metal in glass and successfully explained the high degree of light polarization and the colors obtained on the basis of respective surface plasmon resonances of the conduction electrons in the asymmetric silver metal particles, as was mentioned in Chapter 2. In particular, they explained how the location of the absorption bands

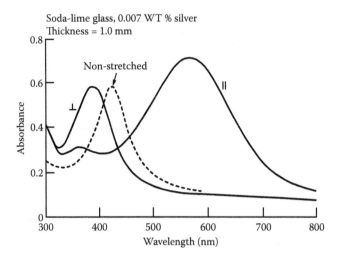

Soda-lime glass, 0.007 WT % silver
Thickness = 1.0 mm

FIGURE 9.1 Wavelength dependence of the absorption changes when silver particles are stretched. The greater the elongation, the more the parallel absorption peak moves to longer wavelengths and the perpendicular absorption peak moves to shorter wavelengths. The spherical (nonelongated) particle absorbs most strongly near 400 nm. (From N.F. Borrelli and T.P. Seward, *New Glass*, 27(1), Serial 104, March 2012; T.P. Seward, *Proc. SPIE* 464, *Sol. State Cont. Dev.*, 1984.[8,12])

depended on the aspect ratios of elongated metal particles and the orientation of a polarized light field with respect to the elongation axis of the commonly aligned particles. Figure 9.1 illustrates this effect. The particle aspect ratios required to produce absorption maxima at various wavelengths can be estimated using equations given in Bohren and Huffman,[5] as is described in Section 9.2.2.

In all these cases and the ones that follow, we refer to elongation as a "redraw." In Appendix 9A we give a brief description of this process as a reference to what is involved.

Stookey and Araujo's work initiated extensive studies of these phenomena at Corning, much of which has been reported in the open literature. For example, it was demonstrated that strong resonance absorption could be produced using copper or gold nanoparticles, but the degree of polarization was limited by the contribution to the dielectric function of interband electronic transitions. This tends to detract from the desired conduction electron resonance condition. The optimum condition obtained from a material that essentially produces a free-electron-like dielectric function that is Drude-like[5] in form (see Section 2.5.1). This is why base metals such as lead and bismuth do not exhibit such resonance and consequently perform poorly as polarizers. There was also a demonstration of a weak polarizing effect in stretched photochromic glass in the UV-darkened state.[6] (See also Chapter 5 for more on this effect.)

9.2 NOVEL PROCESS DEVELOPMENT

As we will see in later sections, the new idea that was proposed for making a polarizing glass was based on the Ag halide- and Cu halide-based photochromic glass

---

TABLE 9.1
Composition of Polarizing
Photochromic Glass

mol%	8111
SiO_2	61.76
Al_2O_3	4.00
B_2O_3	17.19
Li_2O	3.99
Na_2O	4.34
K_2O	4.00
CuO	0.008
Sb_2O_3	
Ag	0.147
TiO_2	1.87
ZrO_2	2.67
PbO	
Cl-	0.451
Br-	0.200

Source: K. Lo and D.A. Nolan, U.S. Patent
No. 4,282,022, Aug. 1982.[2]

---

systems. A representative base glass composition for the commercial polarizing glass is listed in Table 9.1. Elongation of such glasses under high stresses produced an elongation of the liquid metal halide phase and produces a weak polarizing effect when in the darkened state by its photochromic property, as was discussed in Chapter 5.

The novel viewpoint taken by Corning was that the elongated silver halide particles suspended in a glass matrix could be thermochemically reduced to the metallic state. Borrelli and Lo and Lo and Nolan[1,2] showed that if the silver halide particles were elongated and aligned by a suitable mechanical process prior to the reduction step, elongated silver particles resulted after reduction, providing significant light polarizing effects (contrast ratio and transmission). That is to say, a much superior finding than that originally reported by Stookey and Araujo. Moreover, as we will see, this had the significant advantage of being achieved at much lower stresses.[4]

9.2.1 DESCRIPTION OF THE NEW METHOD

Two practical methods for elongating the particles, drawing and/or extrusion at a temperature above the strain point temperature, and two types of particulate materials, either a metal or a secondary phase that can be thermochemically reduced to metal, emerged from these studies. But to understand the difficulties in developing a commercial product as faced by Land, Stookey, and Araujo, it is important to look at the polarization principles and the particle elongation process in some detail.[8]

III A principles: When glasses containing a dispersion of second-phase particles are elongated under a tensile load, the second-phase particles tend to elongate into needlelike (ellipsoidal) shapes aligned along the tensile axis. Such an arrangement of particles can give rise to unusual optical properties, including light polarization and birefringence. If the second phase consists of small, elongated, light-absorbing particles of the refractive index different than that of the matrix glass, polarization effects result.[9] This follows because the amount of light absorbed by the particles is proportional to the square of the electric field developed within the particles. This internal field depends on both the applied field and the electrical polarizability of the particle; however, the polarizability of nonspherical particles depends on the direction of the applied field. It is this difference in polarizability that gives rise to the difference in absorption, and hence, produces a light-polarizing material.

For cases where the absorbing particles' dimensions are comparable to or smaller than the wavelength of light, the degree of polarization can be calculated using light-scattering theory, provided one knows the particle shape and dimensions and the optical constants of the metal involved (see, for example, Van de Hulst[9]).

III B process: If the second phase particles are nonspherical as made (have one axis longer than the others), the particles can be aligned by the flow of the glass. On the other hand, if the particles are spherical or approximately so, the flow process must elongate the particles as well as orient them. In this case the requirements are that (1) the particulate material be deformable (either a liquid, a molten glass phase, or a deformable crystal) and (2) the elongating glass body (matrix) must impose on the particle sufficiently high deformation forces to overcome the surface tension forces tending to maintain the spherical shape and to deform the particle (via viscous flow if it is a liquid or glass, or via dislocation or vacancy motion if it is a crystal).

III C surface tension and elongation forces: Experience has shown that much greater redraw stresses (often in excess of 10,000 psi) are required to elongate silver metal particles of the size needed for good polarization than are required for elongating silver halide particles of comparably useful sizes (a few thousand psi). This is of particular concern for the high degree of elongation needed for NIR polarization and is the primary reason that Polarcor manufacturing uses a two-step process: first precipitate and elongate silver halide particles, then chemically reduce the silver halide to silver metal in a hydrogen gas environment. The requirement for high redraw stress and the difference in stresses required to elongate silver metal and silver halide can be explained in terms of surface tension.

Surface tension (interfacial energy) tends to minimize the surface area of a liquid (or molten glass) droplet, causing it to achieve spherical shape. It also produces an internal pressure within the droplet, given by $P = 2\gamma/r$, where γ is the surface tension and r the radius of the sphere. If the droplet is distorted from its spherical shape, the pressure beneath any point on the surface will be given by $P = 2\gamma(1/r_1 + 1/r_2)$, where r_1 and r_2 are the radii characterizing the curvature of the surface at that point.

A droplet of silver halide, such as found in photochromic glasses and Polarcor products, has a surface tension of about 150 erg/cm^2 with a radius of about 10 nm. According to the pressure equation given above, $(P = 2\gamma/r)$, this would correspond to an internal pressure of about 4500 psi. Therefore, it is understandable that viscous forces of this order of magnitude are needed to significantly elongate such particles by redraw or other techniques.

The air-metal surface energy of noble metals is about an order of magnitude higher than that for silver halides, so even allowing for the metal to wet the glass, the glass-metal interfacial tension is likely still considerably more than that for silver halide, and therefore, correspondingly higher deformation forces are required to deform a similarly sized particle. Another factor favoring the elongation of silver halide over silver metal is that at the deformation temperatures used in drawing or extrusion (temperatures near the softening point of the glass) the silver halide particles are fluid droplets and will elongate (deform) in proportion to the elongation (deformation) of the glass matrix. Metal particles, on the other hand, are in the solid crystalline state and must elongate via dislocation motion or atomic diffusion, thus they elongate more slowly than the host glass.

Figures 9.2 and 9.3 show absorption bands developed in the drawn glasses containing silver and hydrogen-reduced silver halide, respectively. The higher draw stresses required for the silver particles are also shown. Since 1000 psi is often considered a practical maximum engineering design stress level to avoid breakage, the silver-halide-containing material has a clear manufacturing advantage.

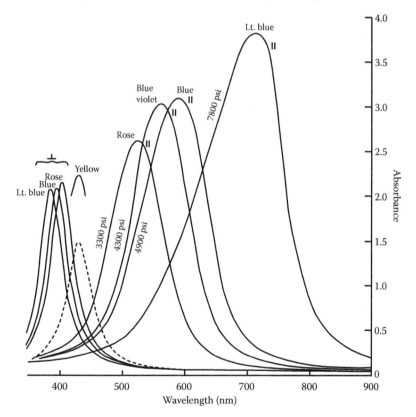

FIGURE 9.2 Wavelength dependence of the absorption of elongated silver metal particles in a soda-lime-silica glass as a function of drawing stress, which are noted on the curves; higher stresses produce greater elongation. The colors noted are those seen when two similar glass samples are crossed and viewed in white light where *t* indicates glass thickness. (From N.F. Borrelli and T.P. Seward, *New Glass*, 27(1), Serial 104, March 2012.[8])

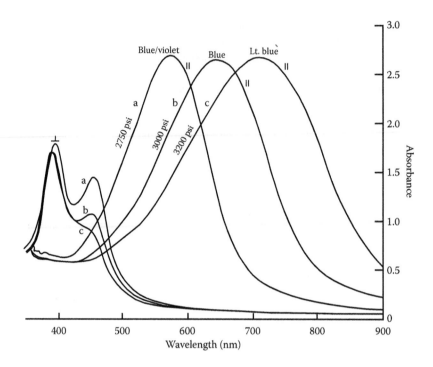

FIGURE 9.3 Wavelength dependence of absorption of elongated silver metal particles in a hydrogen-reduced silver-halide-containing borosilicate glass. The colors noted are those seen when two similar glass samples are crossed and viewed in white light. The smaller absorption peaks near 470 nm are presumed due to small spherical silver particles also produced during the reduction process. (From N.F. Borrelli and T.P. Seward, *New Glass*, 27(1), Serial 104, March 2012.[8])

The hydrogen-reduced metal halide approach has another inherent advantage: The amount of light absorption and contrast (ratio of parallel to perpendicular absorption) can be controlled by the degree of hydrogen reduction, as shown in Figure 9.4.

9.2.2 THE THEORY

According to light-scattering theory, if the metal particles have dimensions that are small compared to the wavelength of the light and are spaced sufficiently far apart to be noninteracting, the absorption cross section per particle may be written in terms of the complex dielectric constant $\varepsilon_c = \varepsilon_1 - i\varepsilon_2$ of the particle material[5,9]

$$C_{abs} = \left(2\pi V n_0^3 / L^2 \lambda\right) \cdot \left(\varepsilon_2 \Big/ \left\{\left[\varepsilon_1 + n_0^2(1/L - 1)\right]^2 + \varepsilon_2^2\right\}\right) \tag{9.1}$$

where V is the volume of the particle, n_0 is the refractive index of the glass matrix, λ is the wavelength in free space, and L is the electric depolarization factor appropriate

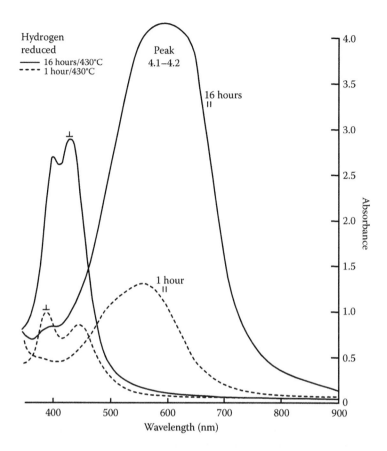

FIGURE 9.4 Wavelength dependence of the absorption of elongated silver metal particles in a hydrogen-reduced silver-halide-containing borosilicate glass as a function of degree of hydrogen reduction. (From N.F. Borrelli and T.P. Seward, *New Glass*, 27(1), Serial 104, March 2012.[8])

for the particle geometry and orientation with respect to the applied field. Whenever one can approximate the particle geometry by a prolate ellipsoid (of axial ratio r = a/b), the L values can be found analytically or in tables.

For the case of absorbing dielectric particles in glass, such as second-phase glassy phase particles, ε_2 can be derived directly from the measured absorption coefficient of that second phase. In the case of metal particles, which are opaque to light, the approach to calculation must depend on published optical constants for the metal in question, or for good conducting metals like silver, gold, and copper that approximate free-electron behavior, one can fit the experimental data to a second- or third-order equation in wavelength. For these materials, a resonant absorption occurs at a wavelength for which the first term in the denominator of Equation 9.1 goes to zero (i.e., when $\varepsilon_1/n_o^2 = (L-1)/L$). This wavelength obviously depends on the optical constants of the metal, the refractive index of the glass, the particle geometry, and the polarization direction of the light (through the depolarization factor L). A wide range of colors can be produced by varying the size and aspect ratio of the particles and the polarization direction of the viewing light.

Based on the free electron theory,[10] the complex dielectric constant of bulk silver metal can be written in the approximate form

$$\varepsilon_c = \varepsilon_1 + i\varepsilon_2 = \varepsilon_0 - A\,\lambda^2 + iB\,\lambda^3 \qquad (9.2)$$

Substituting Equation 9.2 into Equation 9.1,

$$C_{abs} = \left(\lambda^2 \middle/ \left\{\left[\varepsilon_0 - A\lambda^2 + n_0^2(1/L - 1)\right]^2 + B^2\lambda^6\right\}\right) \qquad (9.3)$$

which, when plotted as a function of wavelength, gives a Lorentzian line shape.

It has been found that a good fit with the published literature data for the optical properties of bulk silver can be expressed by the following expressions[8]:

$$\varepsilon_1 = 5 - 55\lambda^2 \qquad (9.4)$$

$$\varepsilon_2 = 0.06 + 27\lambda \exp(-29\lambda^2) + 1.6\lambda^3$$

where λ is expressed in micrometers. The first terms in ε_1 and ε_2 can be interpreted as the wavelength-independent contribution of bound electrons, the second term in ε_2, important only in the ultraviolet (for silver), represents the onset of interband transitions, and the last term in each is the conduction electron contribution that follows Drude-like wavelength dependence. So, in the near-infrared spectral region, $\varepsilon_0 = 5$, $A = 55$, and $B = 1.6$.

The resonant absorption occurs when the first term in the denominator of Equation 9.3 is zero. For a spherical particle of silver in a glass of index 1.5, L = 1/3, putting the resonant absorption at about 416 nm. The wavelength of maximum absorption, λ_{max}, depends on the degree of particle elongation and the orientation of the polarized radiation through the electrical depolarization factor L. For particles of ellipsoidal shape, L ranges from 1/3 to zero for the parallel direction and 1/3 to 1/2 for the perpendicular orientation.

9.2.3 SEM IMAGES

Figure 9.5 shows SEM images of the particles stretched to various degrees corresponding to different resonant wavelengths as noted.

9.2.4 PERFORMANCE AND APPLICATIONS

In the early 1980s, the need for glass near-infrared polarizers became apparent and Corning's Polarcor products were developed.[11-15] Figure 9.6 shows near-infrared

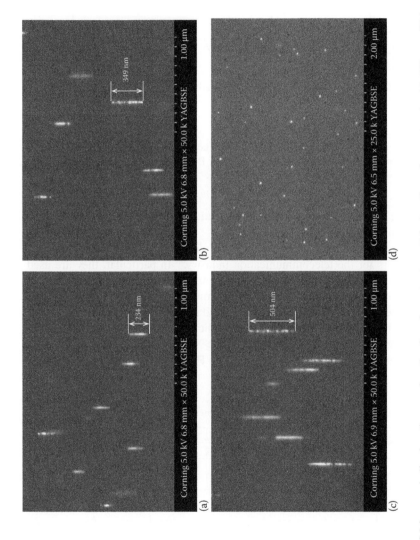

FIGURE 9.5 SEM images indicating the wavelength of peak contrast, particle length, and aspect ratio: (a) 633 nm, 234 nm, 2.7, (b) 900 nm, 349 nm, 3.5, (c) 1500 nm, 671 nm, 77, and (d) the normal image to particle showing a diameter of 87 nm. (From N.F. Borrelli and T.P. Seward, *New Glass*, 27(1), Serial 104, March 2012.[8])

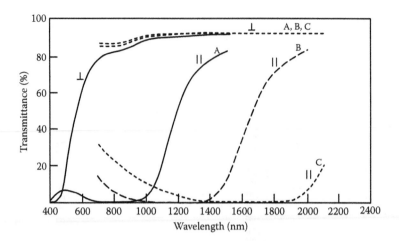

FIGURE 9.6 Near-infrared transmittance spectra of polarizing glasses containing hydrogen-reduced elongated silver halide particles. (From N.F. Borrelli and T.P. Seward, *New Glass*, 27(1), Serial 104, March 2012.[8])

transmittance spectra of polarizing glasses containing hydrogen-reduced elongated silver halide particles.

The major application of such IR polarizing glasses is for polarizing-dependent optical isolators for the telecommunication industry.[16] In particular, they are used for eliminating feedback into the laser diode sources by utilizing a magneto-optic element that rotates the polarization of the incoming light from the laser 45 degrees, whereupon the returning reflected beam is rotated an additional 45 degrees and then is blocked by the polarizer. Typically, for the application greater than 50-db rejection is required. As well, the loss must be minimal; >98% transmission at the operating wavelength including antireflecting coatings (see Figures 9.7 and 9.8).

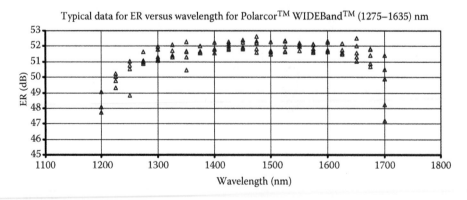

FIGURE 9.7 Contrast in decibels versus wavelength for Polarcor WIDEBand polarizer. (Refer to Polarcor Specification Sheet 000000000094934, Corning Incorporated.)

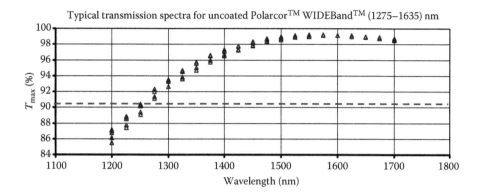

FIGURE 9.8 Transmission versus wavelength, including antireflecting coatings.

9.3   STRETCHED Ag-CONTAINING GLASSES

The original Araujo/Cramer[6] polarizing glass was a stretched Ag nanoparticle in glass. However, the smaller the particle sizes, as discussed above, the greater the restoring forces due to surface tension and the more difficult it is to elongate the particles. As discussed above, the air-metal surface energy of noble metals is about an order of magnitude higher than that for silver halides, therefore correspondingly higher deformation forces are required to deform a similarly sized particle. Nonetheless, a stretched silver nanoparticle can produce reasonably good polarization in the visible portion of the spectrum; the limitation to this spectral region occurs because of the inability to attain large aspect ratios because of the inability to provide sufficient stress. As pointed out in Section 9.2.2, what helps is if the size of the Ag nanoparticle is large. An example is shown in the next section pertaining to a unique glass composition, where a larger size of the Ag nanoparticle could be thermally produced.

9.3.1   PT-DOPED AG-CONTAINING GLASS

Pt-doped Ag-containing glass has the composition listed in Table 9.2. Two things make it unique and distinctive: the very high Ag content and the incorporation of Pt.

TABLE 9.2
Composition of Ag-Based Polarizing Glass

Weight%	ANR
SiO_2	33.96
Al_2O_3	16.65
B_2O_3	13.99
Ag	35.22
Pt	0.1
Cl^-	0.07

Note the extraordinarily high concentration of Ag^+ and yet the glass is observed to be water clear, meaning that the propensity of the Ag to be reduced was entirely eliminated. It was found that from a slight Pt contamination initially thought to be from the crucible produced a glass with the classic Lycurgus cup appearance.[17] For readers not familiar with the term, it refers to old Roman decorative glass that appeared to have one color when looking through it and a different color when seen in diffuse reflection. Figure 9.9 shows a photograph of a glass sample exhibiting this phenomenon for the glass composition listed in Table 9.2 after heating to 650°C.

The effect originates from the fact that when the Ag nanoparticle becomes large enough then the scattering of light becomes as important as the absorption, as will be shown below. The size is gauged relative to the wavelength of light; for a condition where size is $<1/10\lambda$, absorption dominates, whereas when the particle size is $\sim\lambda$, scattering makes a comparable contribution. The former condition for small particles relative to the wavelength of light is referred to as the Raleigh scattering regime and the latter as the Mie scattering regime.[10] We can get some idea how the scattering and absorption cross sections vary with size even in the Raleigh range from the following expressions, where a is the particle radius, and $k = 2\pi\lambda/d\epsilon$ is the complex dielectric constant.

$$\sigma_{abs} = -4(ka)Im\left[\frac{\epsilon-1}{\epsilon+2}\right] \tag{9.5}$$

$$\sigma_{scat} = \left(\frac{8}{3}\right)(ka)^4\left|\frac{\epsilon-1}{\epsilon}+2\right|^2 \tag{9.6}$$

We can readily see how much more rapidly the scattering increases with particle size compared to absorption. Figure 9.10 shows the comparison of the scattered light to the transmitted light for the glass described above. The visual color perception shown in Figure 9.9 comes from the difference in the sharpness of the transmission

FIGURE 9.9 Lycurgus-like effect in the glass of Table 9.2 after 500°C treatment.

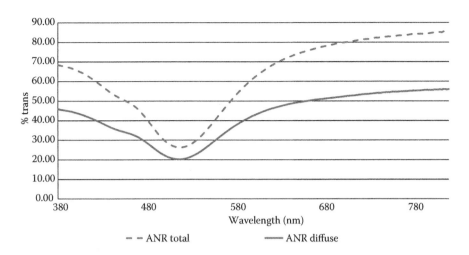

FIGURE 9.10   Total and diffuse transmission of the glass shown in Figure 9.9.

and scattering curves in the 530–630-nm spectral region in Figure 9.10. The more gradual change in slope from the scattering curve accounts for the greenish tint whereas the much sharper change in the transmission curve produces the rose tint. This would have been better shown on a tristimulus color diagram where the relative proportion of the transmission at three wavelengths (red, green, blue) determine the perceived color.

The discussion above points out that the glass can produce large Ag particles and thus will be easier to elongate, but why does one need that high a Ag concentration when only a fraction of a percent of Ag is obtained in the Ag precipitate? The ability to obtain that high a Ag concentration provides an important distinction between the way Ag^+ acts in the glass as compared to alkali ions. Typically when Ag is added to the glass and there are nonbridging oxygens, there is a propensity for the Ag^+ to be reduced to Ag^0. To minimize the nonbridging oxygens, the glass is made to have the alkali content equal to the alumina (also discussed in Chapter 7). However, if the Ag acts as an alkali (ionic) then the condition cannot be maintained. This may point to a different nature of the way Ag bonds into the network, but more importantly, that the nature of this bonding is itself Ag-concentration-dependent.

9.3.2   Pt-Initiated Nucleating Mechanism

The theory of nucleation and growth is a well-studied phenomenon.[18] In general, it follows an approach that breaks the growth of a phase into two stages. The first is the nucleation stage where the nuclei form, whatever they happen to be. It could be a metal particle, such as Ag and Au, or an ion like Ti, or Zr, which seeds a microphase separation, or a spontaneous local enrichment producing a microphase separation. The next phase is the so-called growth phase where from the fixed number of nuclei the particles or crystal phase grows. Clearly, the distinction in where nucleation stops and growth starts is somewhat blurred; nonetheless, the concept is still useful.

One possible simplistic mechanism is posed from a purely redox argument where it is proposed, one is dealing with the following reaction:

$$Pt^0 + 2Ag^+ = Pt^{+2} + 2Ag^0 \tag{9.7}$$

This appears to be an *ad hoc* mechanism since it is not clear how this can be justified on electronegativity alone that Pt should reduce Ag. On the other hand, this equilibrium reaction is likely highly temperature-dependent and perhaps at the heat-treatment temperature the reaction as written makes sense. One does not see any evidence of silver specks in the parent as-melted glass although it does show the characteristic white fluorescence. This fluorescence has been attributed to the $(nAg^0)^+$ molecule.

In another possible mechanism, the Pt particles are conjectured to be the nucleating sites for the growth of Ag particles.

In other words, the reduction of the Ag^+ is nucleated by the Pt^0; the Ag^0 agglomeration is aided by the presence of the Pt particle. If this is the case, then the number of nucleation sites would essentially be fixed by the number of Pt particles. We can clearly see the Pt particles in the as-made glass by SEM, as shown in Figure 9.11. As we see in Figure 9.11, the number of Ag particles seems to far exceed the number of Pt particles, indicating that the mechanism is not likely as proposed.

In the absence of any external agent, one could look at this phenomenon by using an energy diagram similar to what was done in Figure 2.3b. The reason that the Ag stays oxidized is that the energy level corresponding to the reduction to Ag^0 lies above the NBO level, which is the source of the electrons to reduce the Ag, as shown in Figure 9.12. In the case under review, the Pt is added as a $PtCl_4$ solution and could contain as much as 0.07 wt% Cl-ion if it were all retained; that is, not replaced by oxygen during the melting. To the extent that it is retained, one could suggest that it is this nonbridging Cl that supplies the electron to reduce the Ag. Taking the above

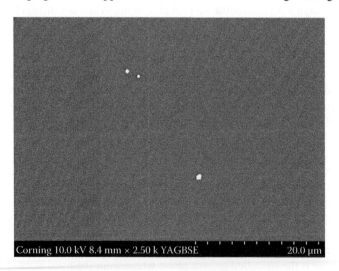

FIGURE 9.11 SEM showing evidence of the Pt inclusion in as-made glass. (From N.F. Borrelli, D.J. McEnroe, and K.R. Rossington; U.S. Patent No. 7,510,989.[19])

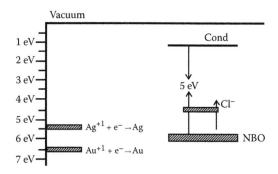

FIGURE 9.12 Schematic energy diagram showing the role of chloride ion.

composition as it stands, then both the chloride and Pt concentrations are ~10^{19}/cm^3. Using the SEM as a guide, one can also very roughly estimate the reduced Ag^0 concentration as the same number.

9.3.3 THE PROCESS

The glass is melted and poured in a mold to form a redraw bar. The bars were thermally treated in the range of 575°C–675°C to strike in the Ag, as shown in Figure 9.13.

The furnace temperature is increased to a temperature in which the glass was soft enough to pull down. For platinum-doped glass, a temperature of 725°C was used for

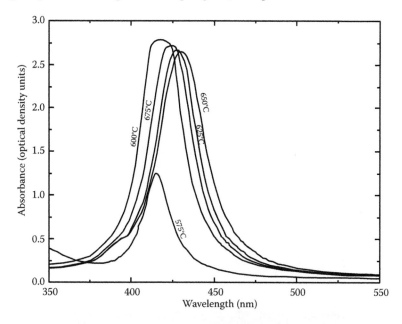

FIGURE 9.13 Strike-in of Ag as a function of temperature. (From N.F. Borrelli, D.J. McEnroe, and K.R. Rossington, U.S. Patent No. 7,510,989; N.F. Borrelli, G.B. Hares, and D. McEnroe, U.S. Patent No. 7,468,148.[19,20])

drawing the glass. Once the glass was initially pulled down, the down feed, which lowers the glass bar into the furnace at a controlled rate, was started. The feed rate of lowering the glass down was set at 13 mm/min. The tractor unit is comprised of two motor-driven belts opposing each other and rotating in opposite directions so the motion through the belts is downward. The distance between the belts can be set so that the glass being drawn through can be grasped by the belts. Figure 9.14a shows the redraw apparatus and Figure 9.14b shows the samples after the various stages.

Figure 9.15 shows the SEM micrograph of the polarizing strip in the draw direction. The aspect ratio of the Ag particle is hard to discern but as we will see for polarizing behavior in the visible region, the aspect ratio need only be ~2–3.

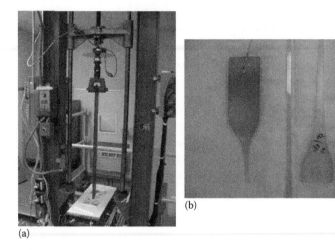

(a)

(b)

FIGURE 9.14 (a) Redraw apparatus, and (b) samples taken from the top of the redraw bar, the strip, and the gob (the part that is grabbed and pulled).

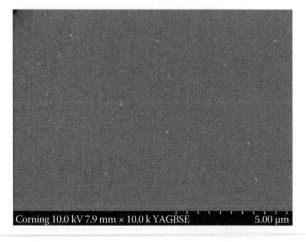

Corning 10.0 kV 7.9 mm × 10.0 k YAGBSE   5.00 μm

FIGURE 9.15 An SEM of a redrawn polarizing bar parallel to the stretch direction showing the slightly elongated Ag particles.

9.3.4 POLARIZED GLASS RESULTS

Figure 9.16a,b show a sampling of the Pt-nucleated Ag polarizing glass spectral performance.[19,20] Figure 9.17 is a representative set of achieved contrast ratio in decibels defined as contrast (db) = 10log (T_{max}/T_{min}), where T is the transmission in the respective polarization direction parallel and perpendicular to the elongation direction,

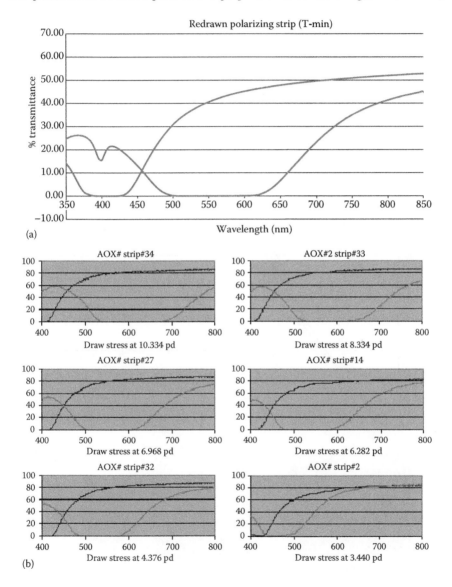

(a)

(b)

FIGURE 9.16 (a) Polarizing performance of the stretched Pt-doped Ag glass. (b) Curves showing how the drawing stress is used to move the wavelength of the peak contrast. (From N.F. Borrelli, D.J. McEnroe, and K.R. Rossington, U.S. Patent No. 7,510,989; N.F. Borrelli, G.B. Hares, and D. McEnroe, U.S. Patent No. 7,468,148.[19,20])

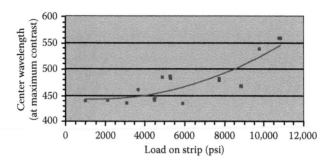

FIGURE 9.17 Data from Figure 9.16b showing quantitatively how the peak contrast center wavelength changes with applied drawing stress. (From N.F. Borrelli, D.J. McEnroe, and K.R. Rossington, U.S. Patent No. 7,510,989; N.F. Borrelli, G.B. Hares, and D. McEnroe, U.S. Patent No. 7,468,148.[19,20])

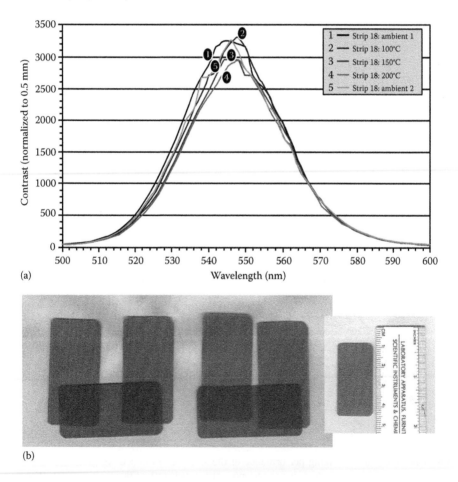

FIGURE 9.18   (a) Polarization contrast versus temperature, and (b) photo of the commercial polarizer.

respectively. For typical LCD display applications, the required contrast is 20 db. Finally, Figure 9.18 shows a correlation of the stress applied and the position of the peak contrast (termed center wavelength). This all comes back to Equation 9.1, where the depolarization ratio L is determined by the elongation aspect ratio, which in turn determines the wavelength of the surface plasmon ratio.

The contrast versus temperature and photo of the actual polarizer prototype production piece is shown in Figure 9.18. The intended commercial use was a replacement of the currently used plastic polarizers in LCD video projectors with the advantage of being able to withstand much higher light intensity and with longer life. These would be the polarizers for the green and red channel. The ability to make an effective blue (460 nm) polarizer is limited by the strong absorption at that wavelength.

APPENDIX 9A: THE METHOD OF HIGH TEMPERATURE REDRAW

Throughout this chapter there are many references to the elongation of a glass to ultimately produce the polarizing property. This process is more often than not referred to as "redraw" but it can be utilized in different ways depending on the effect to be produced. For example it can be used for glass, forming from simply conformally reducing the size of an object like tubing, or an optical waveguide. In this case we use the shear stress applied by the viscous glass to physically elongate the softer Ag halide particle.

The starting blank is called a *preform*—a rod, tube, rectangular bar, sheet, or other desired shape (see Figure 9.14). Generally, the redraw is done at temperatures near the softening point, but depending on temperature and velocities, low or high tensile stresses occur during the redraw. Elongation of a preshaped blank is accomplished by feeding it into a furnace (hot zone) at one rate and pulling it out at a faster rate (see Figure 9A.1). A photo of a laboratory redraw is shown in Figure 9.14.

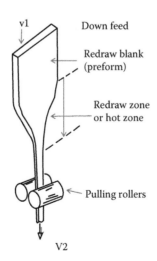

FIGURE 9A.1   A schematic of a typical redraw used to elongate a feature in the glass.

REFERENCES

1. N.F. Borrelli and K. Lo, U.S. Patent No. 4,304,584, December 1981.
2. K. Lo and D.A. Nolan, U.S. Patent No. 4,282,022, August 1982.
3. E.H. Land, U.S. Patent No. 2,139,816, 1928.
4. S.D. Stookey and R.J. Araujo, *Appl. Opt.* 7, 777, 1968.
5. C.F. Bohren and D.R. Huffman, *Absorption and Scattering of Light by Small Particles*, John Wiley & Sons, New York, 1983.
6. R.J. Araujo, W.H. Cramer, and S.D. Stookey, U.S. Patent No. 3,540, 793, November 1970; T.P. Seward III, *Proc. It Int. Congress on Glass*, Vol. 111, 1977.
7. T.P. Seward, *J. Non-Cryst. Solids*, 15, 487, 1974.
8. N.F. Borrelli and T.P. Seward, *New Glass*, 27(1), Serial 104, March 2012.
9. H.C. Van de Hulst, *Light Scattering by Small Particles*, John Wiley & Sons, New York, 1957.
10. N.F. Mott and H. Jones, *The Theory and Properties of Metals*, Dover Publications, New York, 1958.
11. N.F. Borrelli et al. U.S. Patent No. 4,479,819, October 1984.
12. T.P. Seward, *Proc. SPIE* 464, *Sol. State Cont. Dev.,* 1984.
13. T.P. Seward, D.A. Nolan, and N.F. Borrelli, *Tech. Program Opt. Soc. Am. Annual Mtg, Opt. Soc. Am.* P73, 1982.
14. T.P. Seward and D.A. Nolan, *Tech. Program. Opt. Soc. Am. Annual Mtg, Opt. Soc. Am.* P49, 1982.
15. M.P. Taylor, *New Glass*, 7, 108, 1992.
16. N.F. Borrelli, *Microoptics Technology*, Second Edition, Chapter 8, Marcel Dekker, New York, 2005.
17. D. Williams, *Masterpieces of Classical Art*, British Museum Press, 2009.
18. D.A. Porter and K.E. Easterling, *Phase Transformations in Metals and Alloys*, Second Edition, Taylor & Francis, 1992.
19. N.F. Borrelli, D.J. McEnroe, and K.R. Rossington, U.S. Patent No. 7,510,989, 2009.
20. N.F. Borrelli, G.B. Hares, and D. McEnroe, U.S. Patent No. 7,468,148, 2008.

Index

Page numbers ending in "f" refer to figures. Page numbers ending in "t" refer to tables.

Printed and bound by CPI Group (UK) Ltd, Croydon, CR0 4YY

01/11/2024

01782619-0005